engineering *INSPECTION* manual

by

Henry A. Roy

Chief Inspector,
Manning Maxwell and Moore, Inc.,
Aircraft Products Division

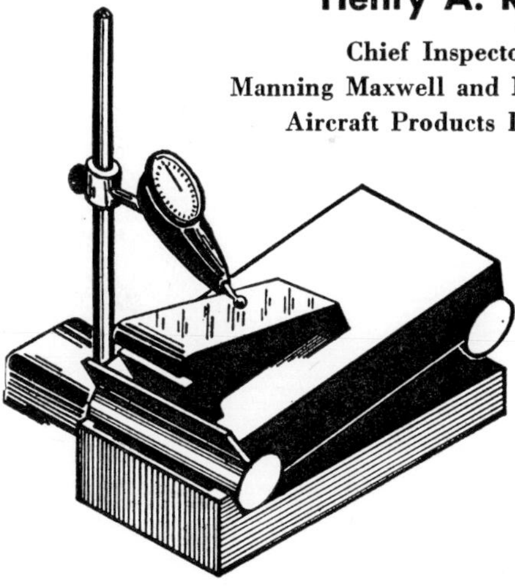

ILLUSTRATIONS BY THE AUTHOR

1956

CHEMICAL PUBLISHING CO., INC.

212 Fifth Avenue New York, N. Y.

CHEMICAL PUBLISHING CO., INC.
212 Fifth Ave., New York, N. Y.

Printed in the United States of America

FOREWORD

This book will help you become a better inspector.

To learn the trade or increase your knowledge on inspection methods, practice those suggested in this book.

Very soon you will develop a knack and ease in making inspection setups.

Your employers will recognize the interest and ability you display in your work.

Your A to Z of advancement in aircraft and machine-tool inspection depends on your ability to do the job better than those around you.

You will have no interest in reading this manual unless you are determined to improve yourself and your position.

The Author

ACKNOWLEDGMENTS

The author wishes to express his appreciation to the following companies for the helpful contributions they have made to this first edition and for permission to include certain information and illustrations in the text:

PRATT AND WHITNEY COMPANY
THE L. S. STARRETT COMPANY
WILSON MECHANICAL INSTRUMENT LABORATORY

CONTENTS

WHY I WROTE THIS BOOK

At no time in the history of the machine-tool and aircraft industries have we been faced with so urgent a need for a greater understanding of inspection procedures as today.

This is true because all industries have reached a crisis in design and mechanical development. If industry is to survive, it has to develop rapidly a better understanding of inspection methods by its inspectors.

It is imperative that any information on the problems of better inspection is brought to the attention of all concerned. This is my reason for writing this book.

Knowing that inspectors often refuse to share their knowledge with others, I shall present as much of my practical, technical, and functional knowledge of inspection as is possible, while continuing my search for better, more efficient methods.

This book is confined to mechanical inspection which I know the most about from first-hand experience.

All inspectors working toward better inspection have different approaches to the problem and it appears that if an attempt were made to include all current techniques and ideas in the same book, the result would be utter confusion and conflict.

My interest in better inspection stems out of the help I derived from it personally. Thoroughly convinced of its necessity and practical value, I was impelled to learn more about the work. During the past twenty-one years, I have collected information and have worked closely with others in order to exchange experiences with them and communicate to them some of my practical findings.

I know from experience and observation how few inspectors live in their work, how much talent is wasted, and how often an inspector does not understand his obligation.

This situation is due to poor training and supervision and to the lack of good human relations between the chief inspector and his department personnel.

An inspector—whether he is doing visual, gaging, first-piece, or final inspection—should be periodically interviewed by the chief inspector. In my experience, if the chief inspector arranges meetings with his inspectors either singly or in a group, he will not only obtain some valuable ideas from them, but will also be able to judge which members of his department are most valuable to the company.

My belief and faith in the importance of good inspection contributed to the development of new inspection and quality-control organizations all over the United States.

I have given only the practical aspects and the results of inspection work, thinking that these will be of the most interest and value.

I have also tried to present some of the results of recent research. I believe that a better understanding of inspection, its functions and theories, will lead to increased efficiency and closer cooperation between manufacturing and inspection.

Although QUALITY must be built in—and cannot be inspected in, carelessness, incapacity, or neglect in inspection is unforgivable.

INSPECTOR'S CREED

I will:

Intelligently approach a foreman or supervisor with respect to poor workmanship.

Never reject a piece of work without just cause.

Study a blue print and route sheet of any job before inspecting the parts.

Properly select the right gages before attempting to check the work.

Eliminate from an order any piece or pieces that do not conform to the drawing and cannot be salvaged.

Calculate the preference of one order over another.

Trust my superior's decisions.

Observe and promulgate any suggestions given me that may help inspection.

Report any job that I am not certain of to the proper authority.

AN INTRODUCTION TO INTERCHANGEABLE PARTS

Eli Whitney presented the world with the idea of mass production about the year of 1798, when he demonstrated the interchangeability of identical firearm parts. His gun parts, however, were not interchangeable with parts made by another gun manufacturer. This was due to the fact that they had no common basis of precision measurements.

As far as I know, mass production did not begin until about 1913.

Interchangeable parts are very important in making a better product and it is equally important to gage and inspect these parts properly to assure the interchangeability of one part with another of the same type.

SCALE AND HOOK RULE

Of all measuring tools, the most common and simplest to use is the steel scale or hook rule (see Figure *a*); it is usually 6 inches long, but other lengths are also available.

The scale most popular with inspectors is that of Figure *a*.

This scale or hook rule sometimes bears graduations on both

Figure a

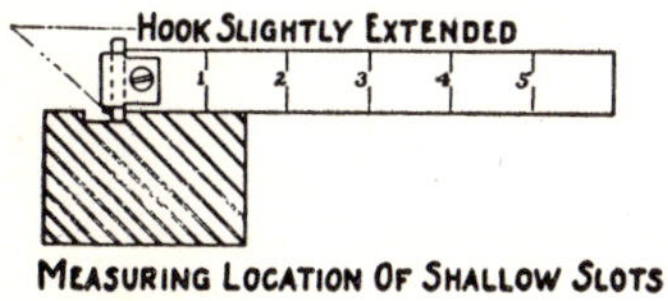

Figure b

Figure c

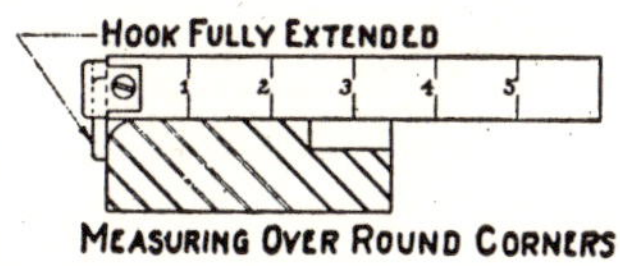

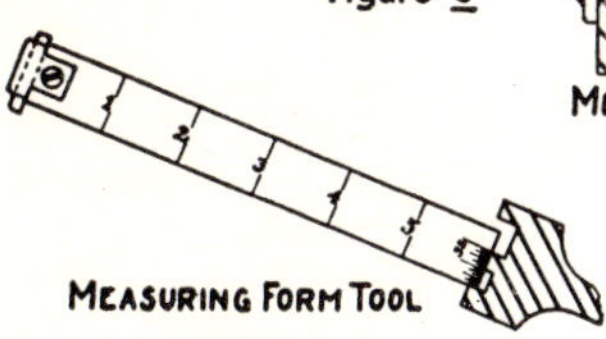

Figure d

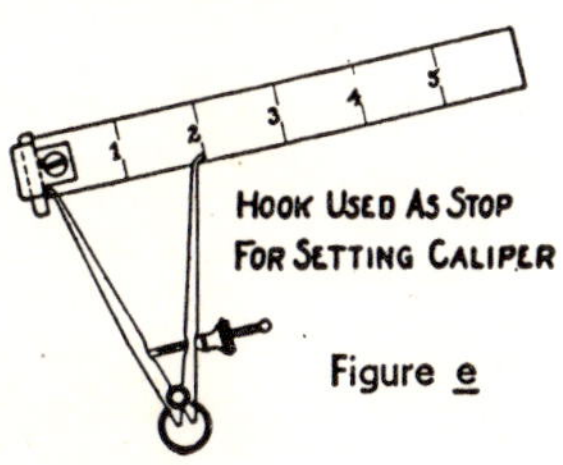

Figure e

Figure f

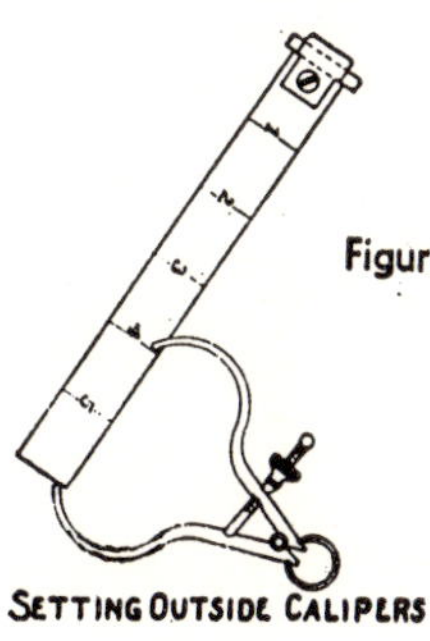

sides and on the ends of each side (see Figure *d*); it is graduated in 1/64, 1/32, 1/16 and 1/8 inch.

The inches may also be divided into tenths, hundredths or in centimeters and millimeters.

The uses of the scale or hook rule are illustrated in Figures *b* to *f*.

COMBINATION SQUARE

A combination square serves a wide variety of purposes.

It consists of a graduated steel blade (or scale), a center head, and a stock, the last two being movable relative to the blade (see Figure *a*).

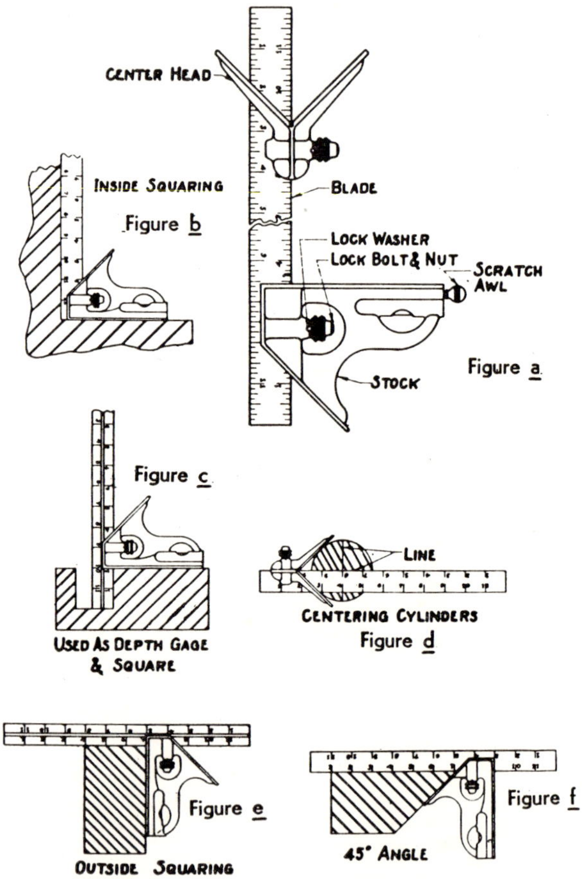

The center head and stock can be adjusted to any position along the blade and locked securely, or one or both may be removed entirely.

Figures *b*, *c* and *e* show some of the uses of the combination square in which only the stock is attached to the blade.

Figure *d* shows how the center head is used to find the center of a cylinder.

Figure *f* shows how the stock and blade are used to check a 45° angle.

PROTRACTOR

The protractor head, commonly called the bevel protractor, is attached to a scale, adjusted to any position, turned to any angle, and locked (see Figure *a*).

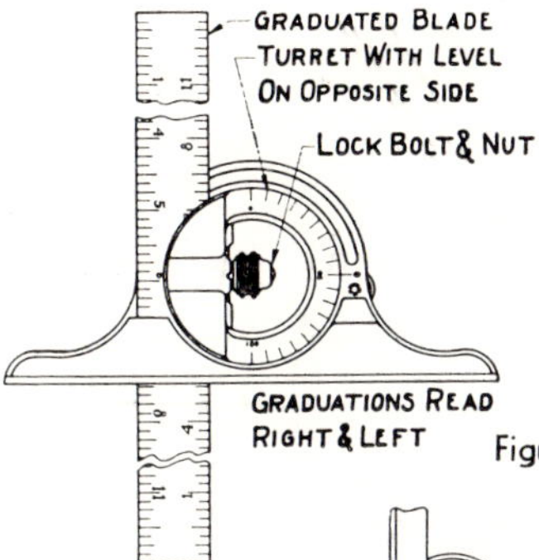

Figure a

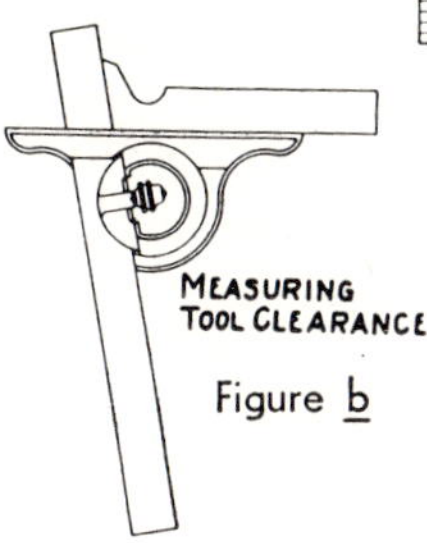

Figure b

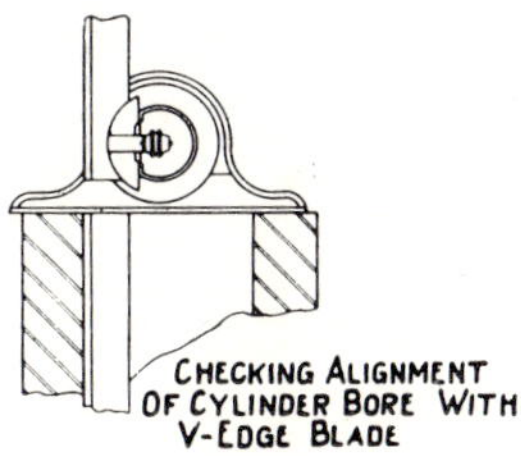

Figure c

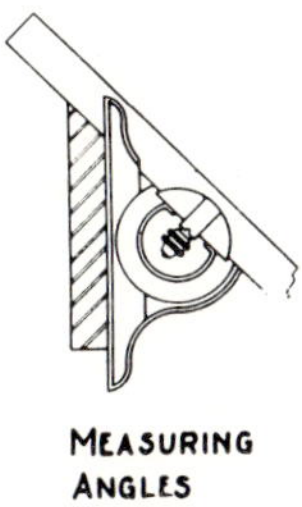

Figure d

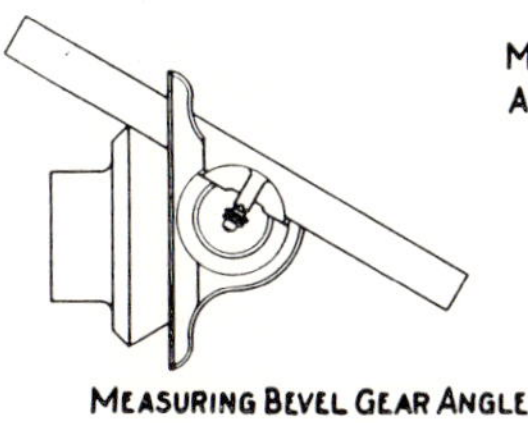

Figure e

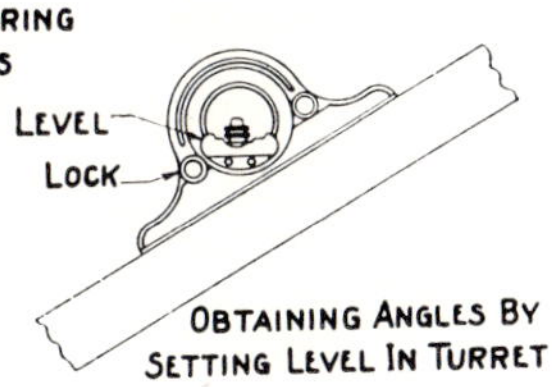

Figure f

The angular graduations usually read from 0 to 180° both ways.

It may be equipped with a spirit level, which is adjustable to show any required degree.

The Figures *b* to *f* show some of the uses of protractors.

OUTSIDE MICROMETER CALIPER

The micrometer caliper—commonly called the "mike" is widely used in inspection.

A very sensitive touch is necessary for accurate measurement.

Different inspectors, measuring the same piece of work with the same mike may obtain different readings. To avoid inaccuracies, many micrometers are equipped with a ratchet on the thimble end (see Figure *a*). This ratchet is adjusted so as to slip when the proper pressure is reached on the anvils. When there is no ratchet, the pressure applied to the work should be such that the work will drag slightly when passed between the anvils. The micrometer shown in Figure *a* has a lock nut to prevent the spindle from moving after it has been set.

The micrometer is used by holding the frame in the palm of the hand, with the third finger resting on the inside of the frame, and rotating the thimble with the thumb and first finger (see Figure *e*).

With large micrometers, it is necessary to use both hands; one to hold the frame and the other to turn the thimble.

It is wrong to hold the thimble and swing the frame around, as this will cause very rapid wear of the threads.

Several of many places where the micrometer can be used are shown in Figures *b, c* and *d*.

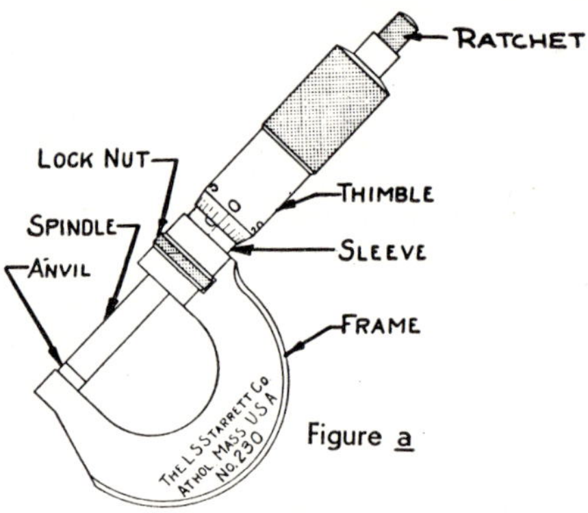

Figure a

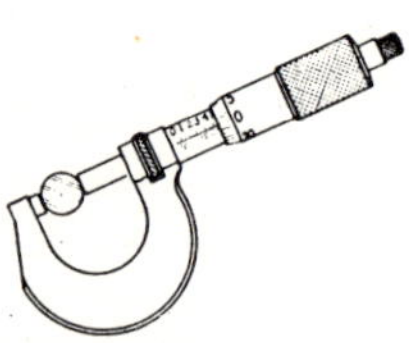

MEASURING DIAMETERS

Figure b

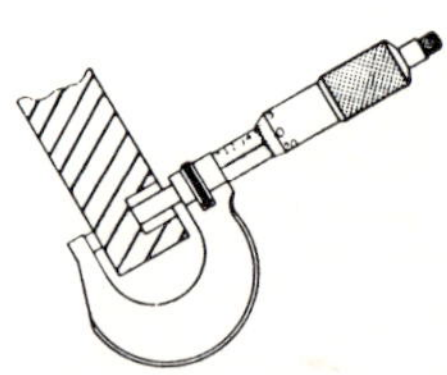

MEASURING IN A SLOT

Figure c

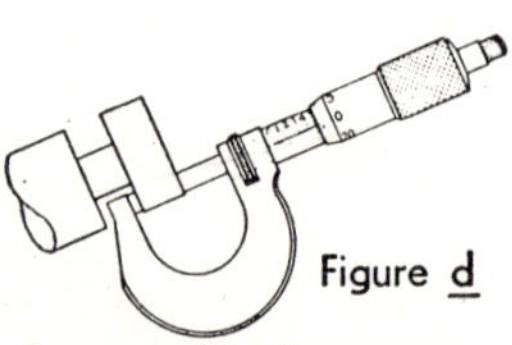

Figure d

SHOWING THE ADVANTAGE OF THE CUT OUT FRAME

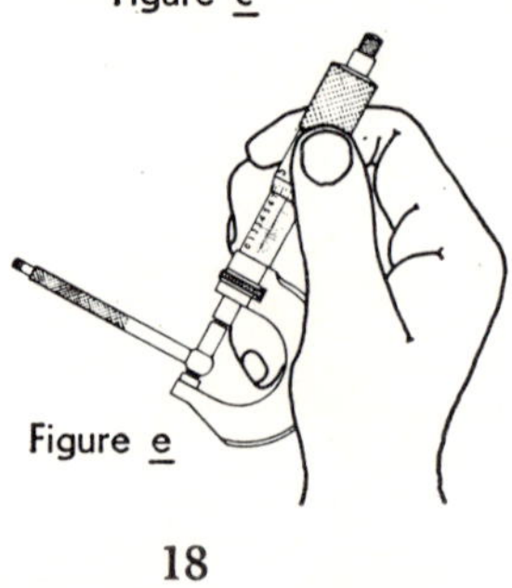

Figure e

MICROMETER DEPTH GAGE

The micrometer depth gage is used to measure the exact depth of a hole or slot (see Figure *a*).

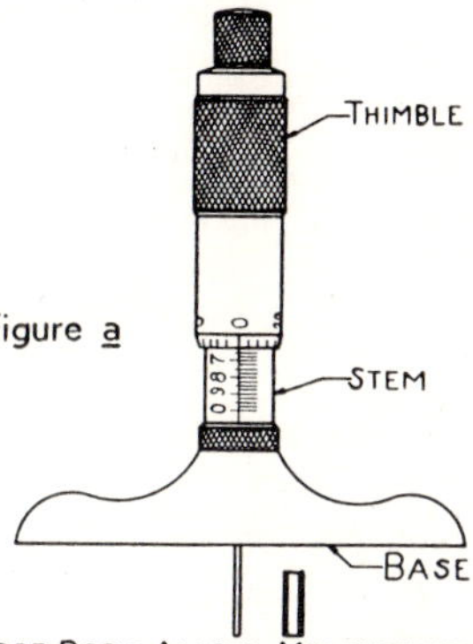

Figure a

THREE RODS ALLOW MEASUREMENTS 0 TO 1", 1" TO 2", & 2" TO 3."

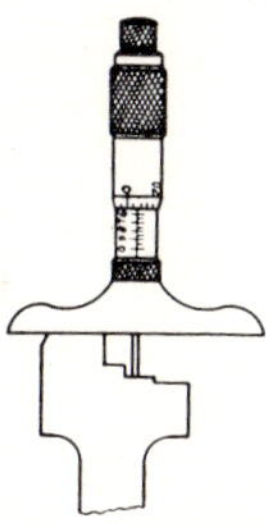

Figure b

FORMED TOOL MEASUREMENTS

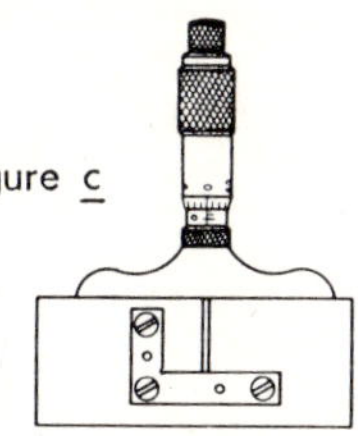

Figure c

USED FOR ACCURATE LOCATING

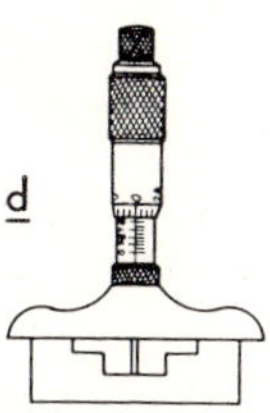

Figure d

DEPTH OF MILLED SLOT

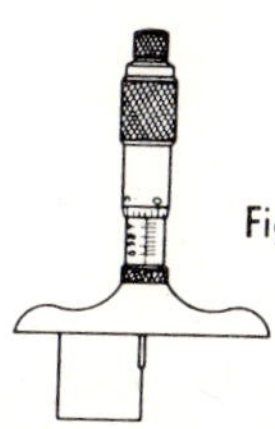

Figure e

MEASURING SHALLOW RECESS

When this instrument is used, the base should be held down firmly and the spindle carefully lowered until it touches the bottom of the hole or slot, (see Figures *b* to *e*).

Very light pressure should be exerted or the base will be lifted off the surface on which it rests and false readings will be obtained.

For accurate readings, a ratchet screw is sometimes provided at the end of the thimble.

INSIDE MICROMETER CALIPER

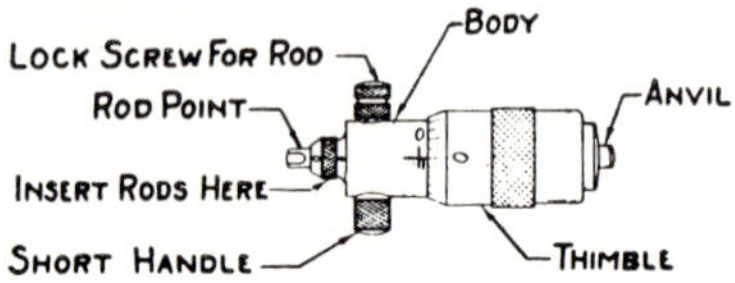

Figure a

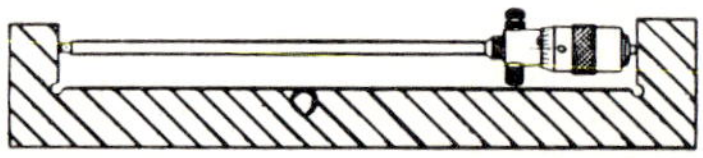

Measuring Length Gauge By Using Long Rods

Figure b

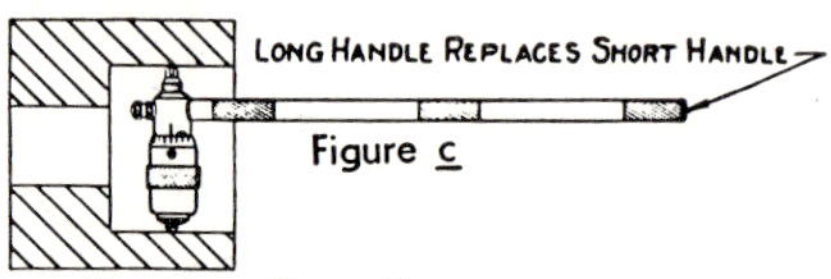

Figure c

Measuring Small Diameters

Figure d

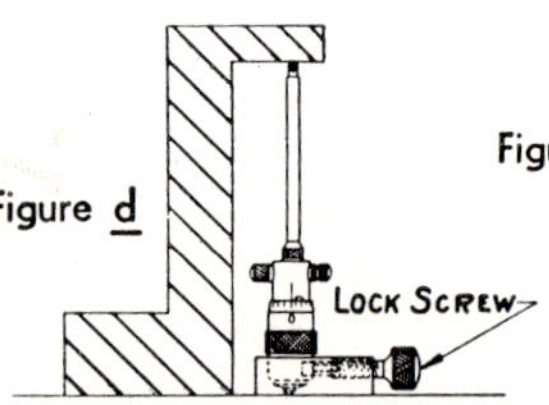

Converted Into Height Gauge

Figure e

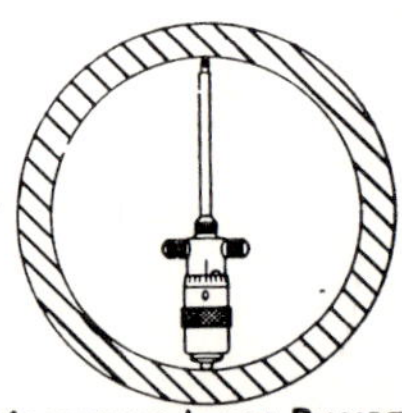

Measuring Large Diameters By Using Long Rods

The inside micrometer caliper is a relative of the ordinary micrometer caliper (see Figure *a*).

It is equipped with extension rods to permit a wide range of measurements.

Select a rod of the proper length and adjust it in the caliper body.

Place the instrument in the hole to be measured (see Figure *e*), with the rod end on one wall, and turn the thimble until its anvil comes into contact with the opposite wall.

Make sure that the instrument is at right angles to the axis of the hole. To accomplish this, move the caliper to the right and left; also up and down to make the proper contact.

Read the full inches from the rods and the thousandths from the barrel and thimble.

For use in a deep hole, an auxiliary handle may be used (see Figure *c*).

A special feel must be developed to use this caliper correctly.

HOW TO READ A MICROMETER CALIPER

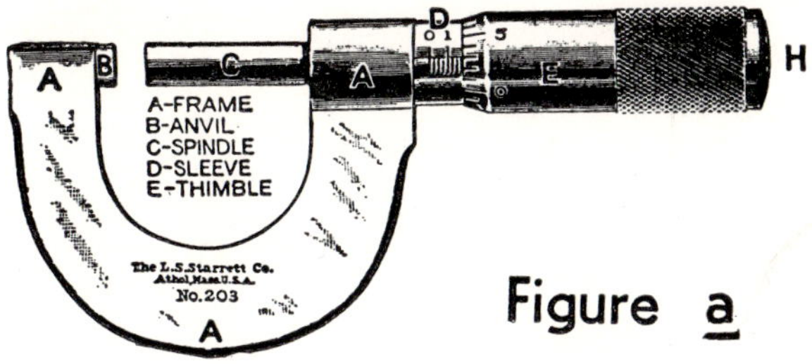

Figure *a*

The spindle C is attached to the thimble E, on the inside, at the point H. The part of the spindle which is concealed within the sleeve and thimble is threaded to fit a nut in the frame A. The frame being held stationary, the thimble E is revolved by the thumb and finger, and the spindle C being attached to the thimble revolves with it and moves through the nut in the frame, approaching or receding from the anvil B. The article to be measured is placed between the anvil B and the spindle C. The measurement of the opening between the anvil and the spindle is shown by the lines and figures on the sleeve D and the thimble E.

The pitch of the screw threads on the concealed part of the spindle is 40 to an inch. One complete revolution of the spindle, therefore, moves it longitudinally one fortieth (or twenty-five thousandths) of an inch. The sleeve D is marked with 40 lines to the inch, corresponding to the number of threads on the spindle. When the caliper is closed, the beveled edge of the thimble coincides with the line marked 0 on the sleeve and the 0 line on the thimble agrees with the horizontal line on the sleeve.

Open the caliper by revolving the thimble one full revolution, or until the 0 line on the thimble again coincides with the horizontal line on the sleeve; the distance between the anvil B and the spindle C is then one fortieth (or 0.025) of an inch, and the beveled edge of the thimble will coincide with the second vertical line on the sleeve. Each vertical line on the sleeve indicates a distance of $\frac{1}{40}$ inch. Every fourth line is made longer than the others, and is numbered 0, 1, 2, 3, etc. Each numbered line indicates a distance of four times $\frac{1}{40}$ inch, or one tenth.

The beveled edge of the thimble is marked in 25 divisions, and every fifth line is numbered from 0 to 25. Rotating the thimble from one of these marks to the next moves the spindle longitudinally $\frac{1}{25}$ of twenty-five thousandths or one-thousandth of an inch. Rotating it two divisions indicates two thousandths, etc. Twenty-five divisions will indicate a complete revolution, 0.025 or $\frac{1}{40}$ inch.

To read the caliper, therefore, multiply the number of vertical divisions visible on the sleeve by 25 and add the number of divisions on the bevel of the thimble, from 0 to the line which coincides with the horizontal line on the sleeve. For example, as the caliper is represented in Figure *a*, there are 7 divisions visible on the sleeve. Multiply this number by 25, and add the number of divisions shown on the bevel of the thimble, 3. The micrometer is open one hundred and seventy-eight thousandths. ($7 \times 25 = 175 + 3 = 178$.)

VERNIER CALIPER

The vernier caliper can be used for measuring both inside and outside surfaces and permits a direct reading between two surfaces to 0.001 inch accuracy (see Figure *a*).

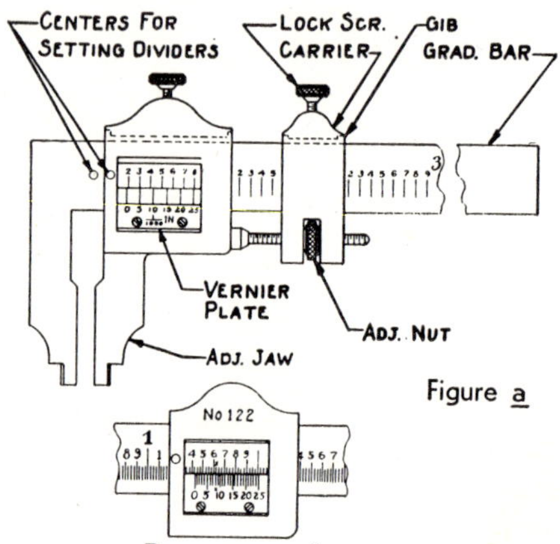

Figure a

READING 1.433"

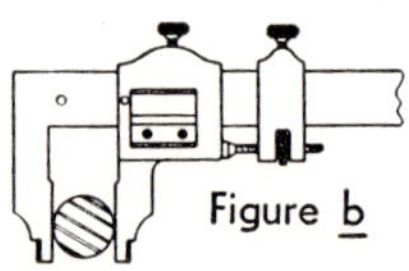

Figure b

MEASURING CYLINDRICAL STOCK

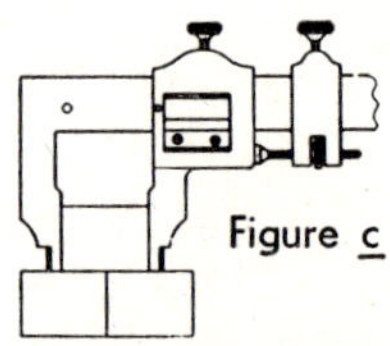

Figure c

OUTSIDE MEASUREMENT OF PUNCH HOLDER

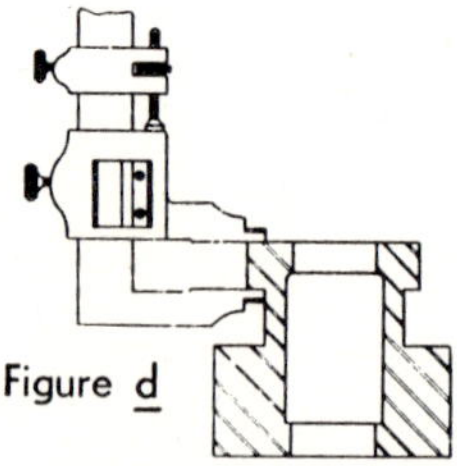

Figure d

LENGTH OF SHOULDER ON TURNED WORK

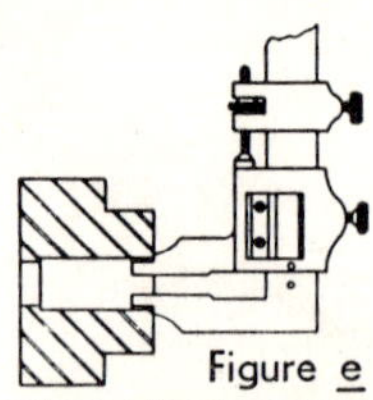

Figure e

TAKING INSIDE MEASUREMENTS

It has a wide measuring range from a pocket size which usually measures 0 to 3 inches to other sizes which measure up to 5 feet.

Outside measurements are taken between the inner faces of the two jaws (see Figures *b, c* and *d*). Internal measurements are taken between the nibs at the outer sides (see Figure *e*).

VERNIER HEIGHT GAGE

The vernier height gage is used for precision measuring and marking vertical distances from a plane surface.

For layout work a scriber is attached to the adjusting jaw and adjusted to the required height by means of the vernier (see Figure *a*).

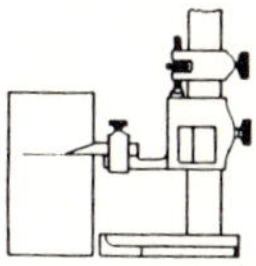

SCRIBING LINE

Figure a

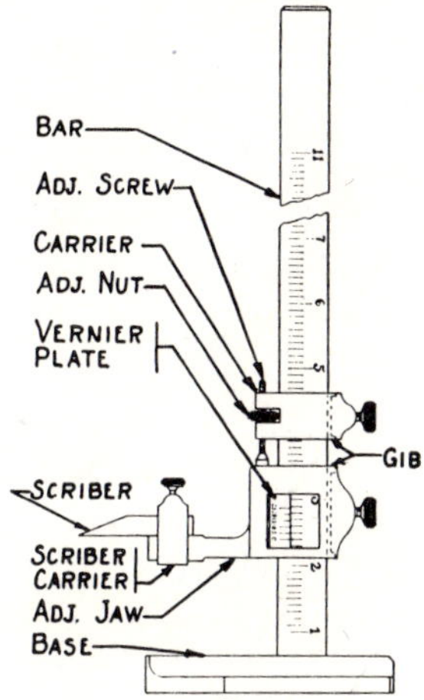

Figure c

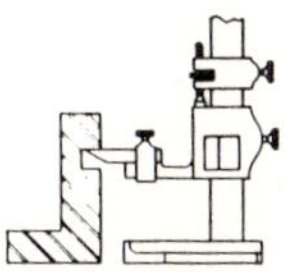

TAKING MEASUREMENTS IN SLOT

Figure b

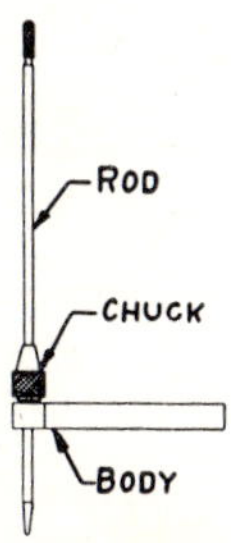

DEPTH GAGE ATTACHMENT

Figure d

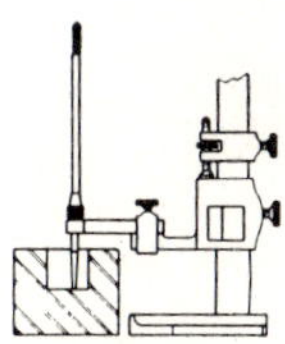

DEPTH GAGE

Figure e

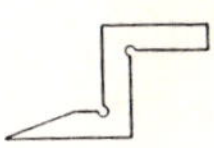

OFFSET SCRIBER

Figure f

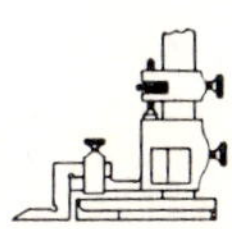

Figure g

OFFSET SCRIBER ENABLES MEASUREMENTS FROM BASE

If lines are required below the height of the gage base, a special offset scriber illustrated in Figures *f* and *g* should be used.

For measuring the depth of a hole, fasten the depth gage attachment, shown in Figure *d*, on the adjusting jaw. First adjust the rod to the top face of the hole and take a reading on the vernier. Then, when the rod is adjusted to the bottom of the hole, the depth from the top face will be the difference between the two readings (see Figure *e*).

HOW TO READ THE VERNIER

The bar of the vernier is graduated in fortieths or 0.025 inch, every fourth division, representing a tenth of an inch, being numbered. On the vernier plate there is a space divided into twenty-five parts and numbered 0, 5, 10, 15, 20, 25. The twenty-five divisions on the vernier occupy the same space as twenty-four divisions on the bar.

The difference between the width of one of the twenty-five spaces on the vernier and one of the twenty-four spaces on the bar is, therefore, 1/25 of 1/40 or 1/1000 inch. If the vernier is set so that its 0 line coincides with the 0 line on the bar, the line to the right of the 0 on the vernier will differ from the line to the right of the 0 on the bar by 1/1000 inch; the second line by 2/10000 inch and so on.

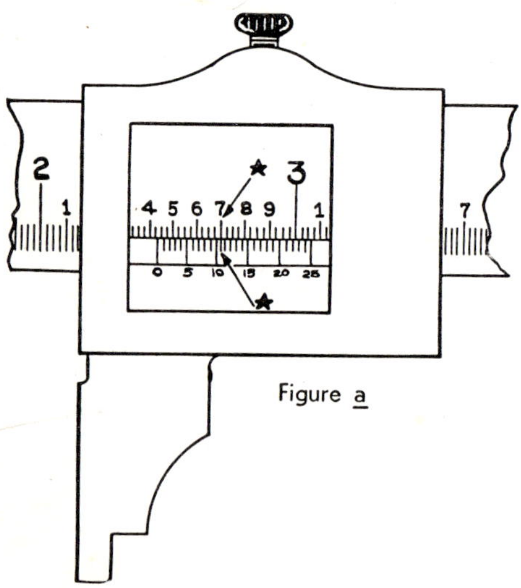

Figure a

The difference will continue to increase 1/1000 inch for each division until the line 25 on the vernier coincides with the line 24 on the bar.

To read the vernier, note how many inches, tenths (or 0.100) and fortieths (or 0.025 the 0 mark on the vernier is from the 0 mark on the bar; then note the number of divisions on the vernier from 0 to a line which exactly coincides with a line on the bar. *Example:* in Figure *a,* the vernier has been moved to the right two and four tenths and one fortieth inches (2.425), as shown on the bar, and the eleventh line on the vernier coincides with a line, as indicated by the stars, on the bar. Eleven-thousandths of an inch are, therefore, to be added to the reading on the bar and the total reading is two and four hundred and thirty-six thousandths inches (2.436).

SURFACE GAGE

The conventional surface gage can be used for layout work or as an indicator stand.

When the scriber is assembled with the surface gage, it is easy to scribe lines on the work (see Figure *a*). As one end of the

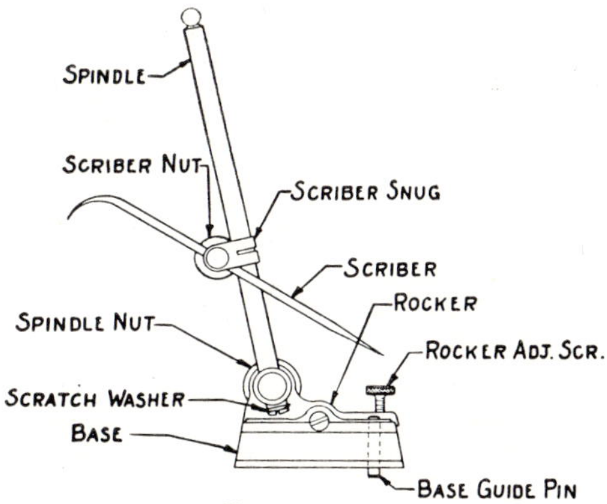

Figure a

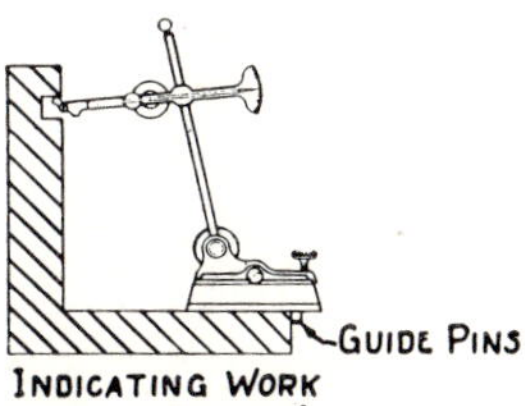

Figure b

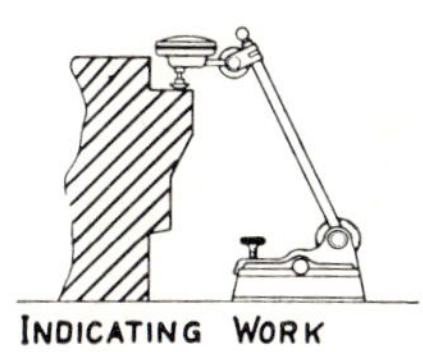

Figure c

Figure d

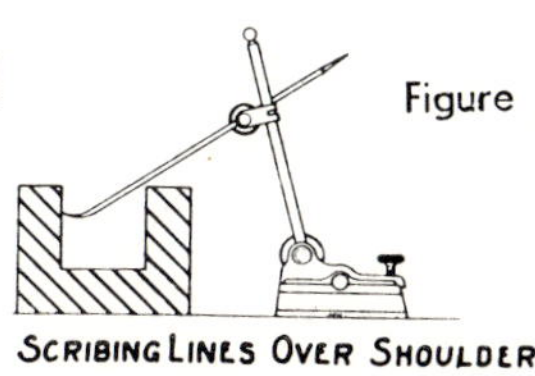

Figure e

scriber is bent, it is also possible to scribe lines in difficult places (see Figures *d* and *e*).

With the scriber removed, several types of indicators can be attached to the spindle for indicating work heights (see Figures *b* and *c*).

UNIVERSAL BEVEL PROTRACTOR

Figure b

The universal bevel protractor is a precision tool designed to check angles accurately to 5 minutes (5′) or 1/12th of a degree (see Figure *a*).

Figure *b* shows some of the combinations that can be obtained with this precision tool.

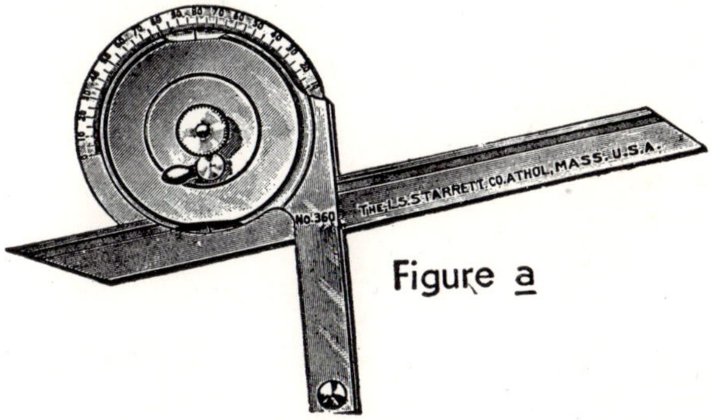

Figure a

How to Read the Bevel Protractor

In Figure *c*, the zero on the vernier scale lies between 52 and 53 on the big protractor dial, indicating 52 whole degrees. Reading from the left, the fourth line of the vernier scale (marked 45) coincides with 70 on the protractor dial as marked by stars.

The reading of the protractor is, therefore, 52 degrees and 45 minutes (52° 45′).

Figure c

THICKNESS OR FEELER GAGE

Figure a

This gage consists of thin, flexible steel strips. Each leaf may be used separately or in combination with others. The leaves are marked with the proper thickness for identification (see Figure *a*).

The thickness gage is used for checking clearances between surfaces (see Figure *b*).

Each strip must be handled with care to avoid kinking or nicking the edges.

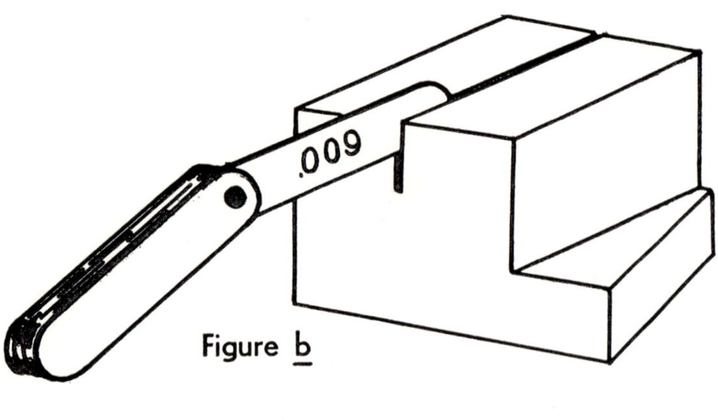

Figure b

SCREW-PITCH GAGE

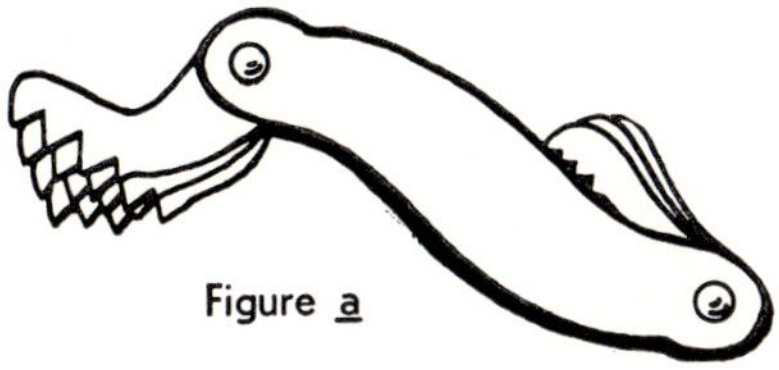

Figure *a*

A screw-pitch gage serves to check or determine the pitch of threads (see Figure *a*).

In use, its successive leaves are placed on a thread of unknown pitch until one of the leaves is found to coincide (see Figures *b* and *c*). The value of pitch is stamped on the side of the leaf.

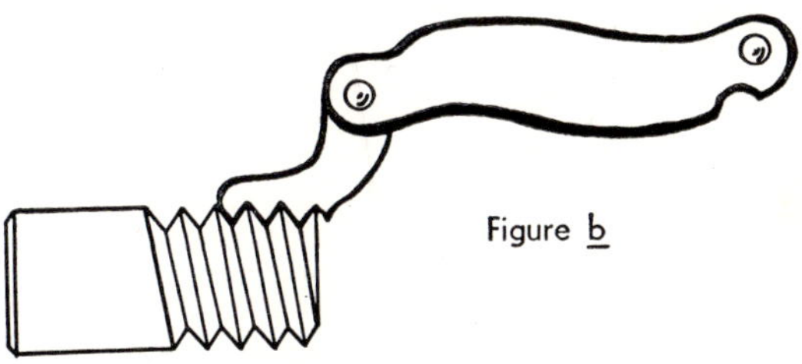

Figure *b*

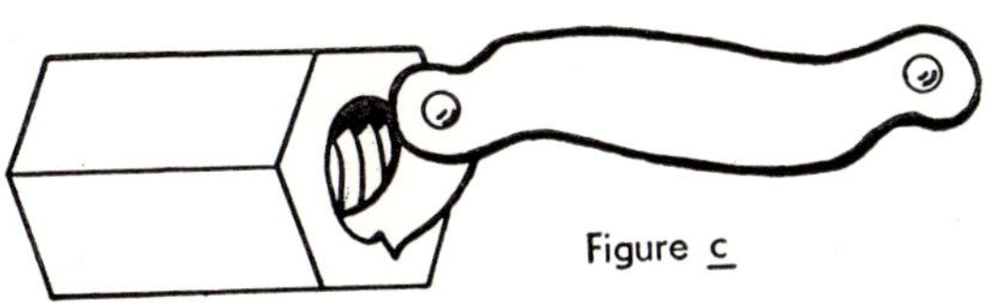

Figure *c*

These gages are available in all standard and metric thread forms.

SMALL-HOLE GAGE

The small-hole gage consists of a small, split, ball-shaped member mounted on the end of a handle.

The ball end is inserted into the hole to be inspected and, by turning the knurled knob on the handle, the split ball will expand and come into contact with the wall of the hole (see Figure *a*).

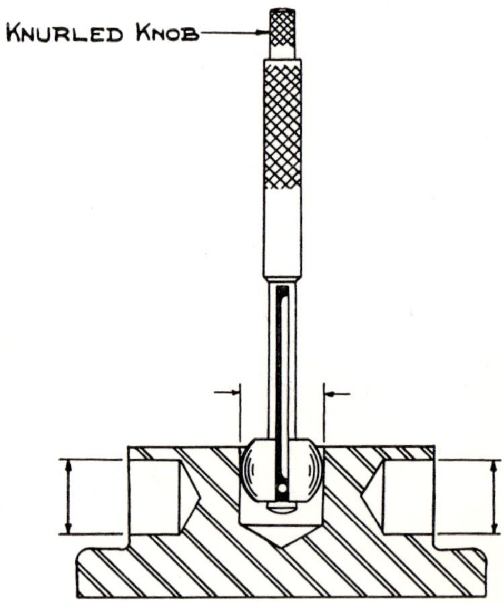

Figure a

When the proper feel is obtained, the gage is removed from the hole whose size is determined by a micrometer-caliper reading taken over the split ball.

TELESCOPING GAGE

The telescoping gage is a T-shaped instrument in which the shaft of the "T" acts as a handle and the telescoping cross arm is used for measuring.

In use, the two arm sections are compressed and placed into the hole. Then they are allowed to expand under the effect of a spring which is between the two arm parts.

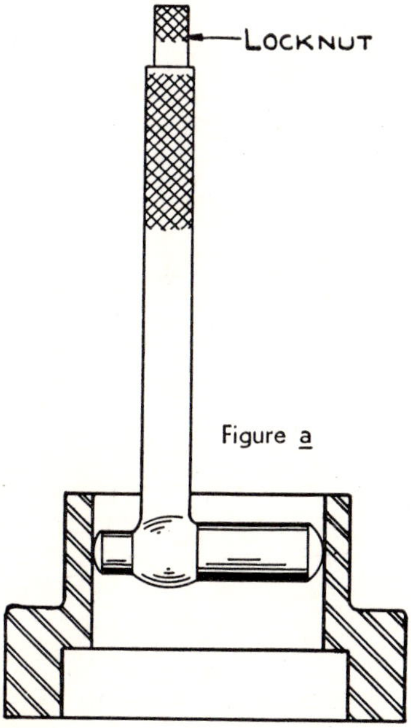

Figure a

At the end of the handle is a lock nut. When the crossarm parts are properly set against the walls of the hole, they are locked in position by turning the lock nut (see Figure *a*).

Then the gage is withdrawn from the hole and the distance across the arm sections is measured with an outside micrometer caliper (see Figure *e* on page 18).

The telescoping gage is available in various sizes.

LOCATING CENTERS

The centers of an ordinary bar stock may be located by scribing lines with a hermaphrodite caliper (see Figure *a*).

Place one of the caliper arms against the outside diameter of the bar and adjust the other arm with the pointer in it to the approximate center of the bar.

Hold the arm that is on the outside of the bar with one hand

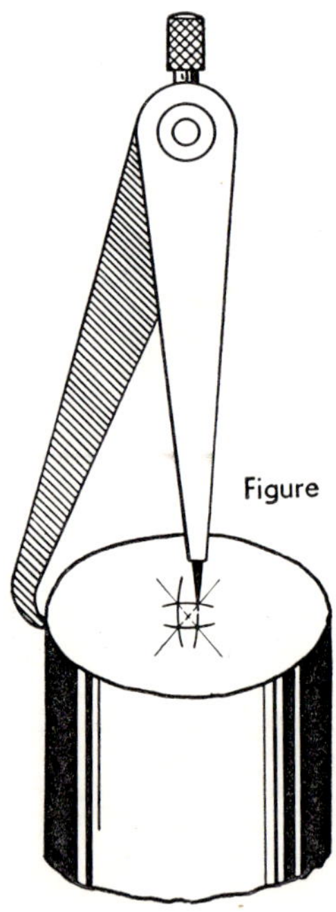

Figure a

and with the other hand, swing the pointed arm from left to right, scribing a line on the bar.

Without disturbing the setting of the caliper, move the outside arm 90° to a new position on the bar and scribe another line. Follow the same procedure until four lines are marked. Where these lines intersect, scribe two more lines. The intersection of the last two lines is the approximate center of the bar.

OUTSIDE AND INSIDE CALIPERS

Calipers are used to measure inside or outside diameters.

Calipers for measuring outside diameters are of the bowed type (see Figure *a*). Calipers for measuring inside diameters have straight legs, with the measuring ends turned out (see Figure *b*).

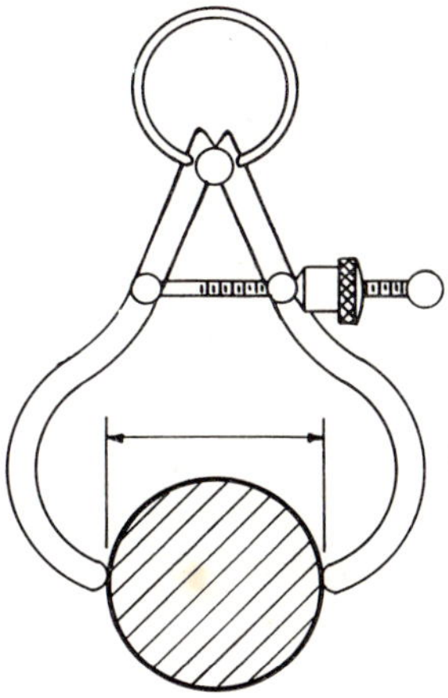

Figure a

Figure b

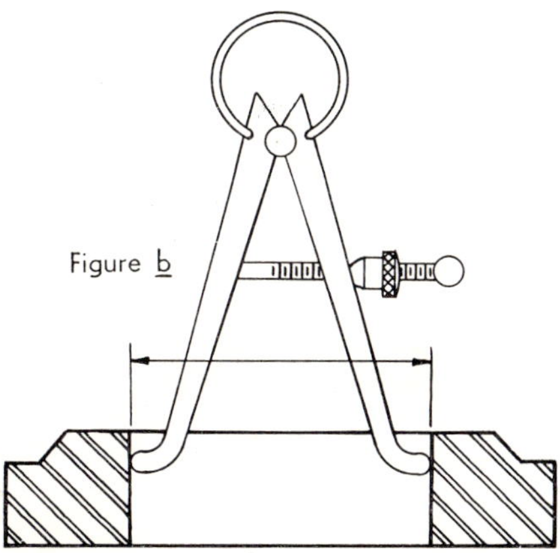

Either of these calipers is used with a steel rule.

To be able to force calipers over or into a work piece, some practice is necessary.

In measuring a piece of work, the legs should pass over it with a slight drag.

When the caliper is adjusted to the work and barely hangs on without falling off, the touch is right.

COMBINATION BEVEL

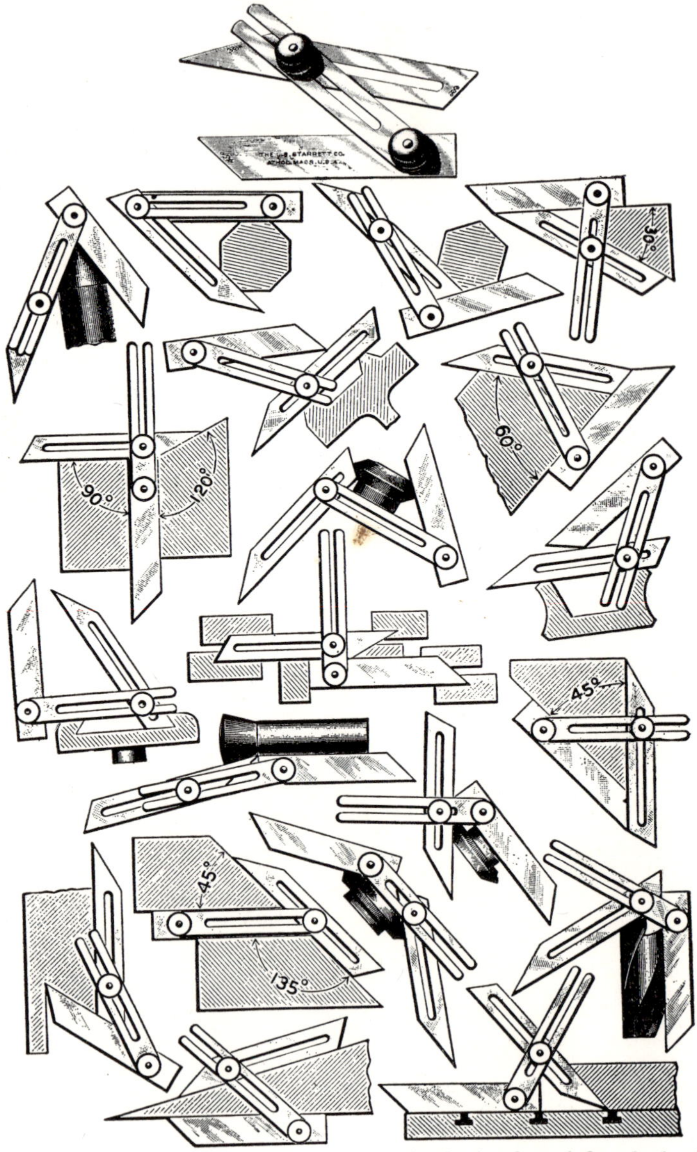

This bevel can be swung to any angle desired and firmly locked in position. It is very useful to an inspector when laying out work or measuring.

The bevel can be preset to any desired position and used as a gage to check many machined pieces.

FILLET OR RADIUS GAGE

This gage, which is also called a concave and convex gage, is equipped with two sets of leaves. The set of leaves at the right side of Figure *a*, is used to check radiuses (see Figures *b* and *c*).

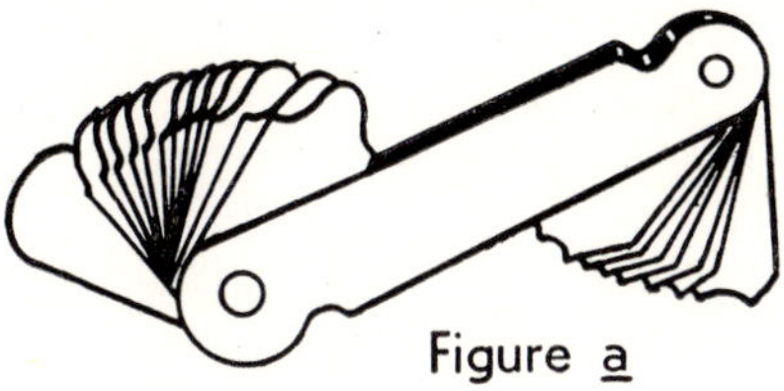

Figure a

The set of leaves at the left side of Figure *a* is used to check fillets (see Figure *d*).

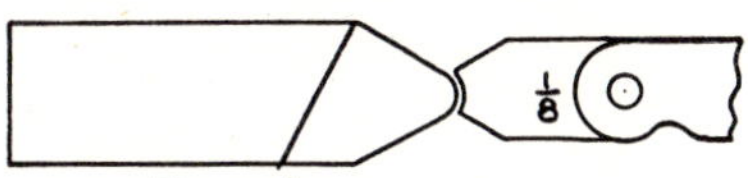

Figure b

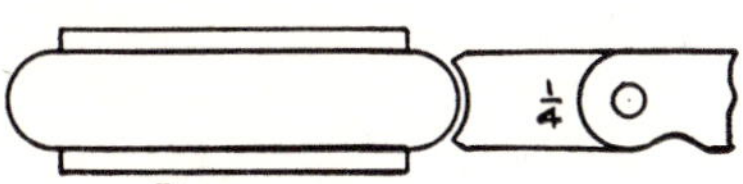

Figure c

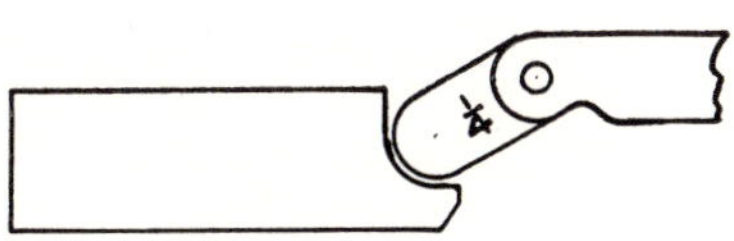

Figure d

It is recommended for machinists, toolmakers and screw-machine operators, as well as pattern makers.

FLUSH-PIN GAGE

A flush-pin gage is used for checking the distance between two surfaces. It consists of a body with a through hole and a pin in the hole which projects from a face of the body a distance equal to the dimension to be gaged when the opposite or indicating end of the pin is flush with the opposite face of the body. The indicating end of the pin, or the adjacent face of the body, has a step of a depth equal to the tolerance on the dimension gaged (see Figure *a*).

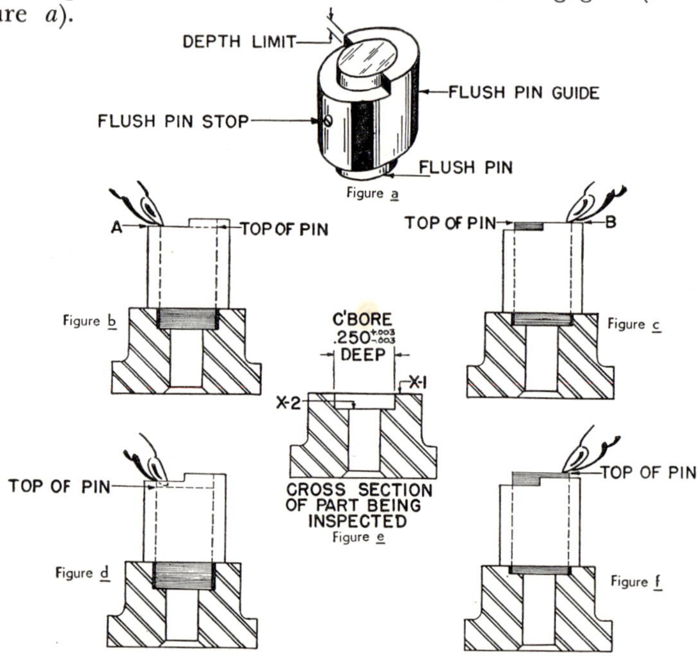

Figure a, Figure b, Figure c, Figure d, Figure e, Figure f

How to Use the Flush-Pin Gage

1. Place the flat side of the gage squarely and firmly on top of surface X-1 (see Figure *e*).
2. Push the pin down until it comes into contact with surface X-2 (see Figure *e*).
3. By passing the fingernail over surface A (Figure *b*) and surface B (Figure *c*), the 0.250 ± 0.003 depth of the C'bore is determined.

Figure *b*—Top of pin flush with surface A; the C'bore is 0.247 inch.

Figure *c*—Top of pin flush with surface B; the C'bore is 0.253 inch deep.

Figure *d*—Top of pin below surface A; the C'bore is too deep.

Figure *f*—Top of pin above surface B; the C'bore is too shallow.
Note: If the top of the pin stops between surfaces A and B, the depth of the C'bore is within the ±0.003 inch allowed.

TAPER-PLUG GAGE

A taper-plug gage is an internal gage for the size control of conical holes. It is similar to a plain cylindrical-plug gage, but its gaging part is tapered (see Figure *a*).

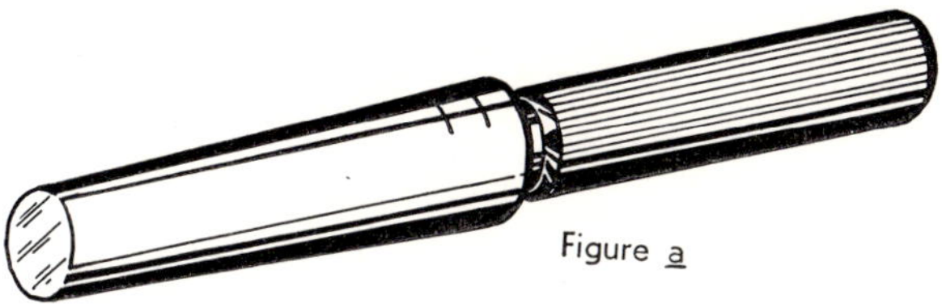

Figure a

Figure b

Figure c

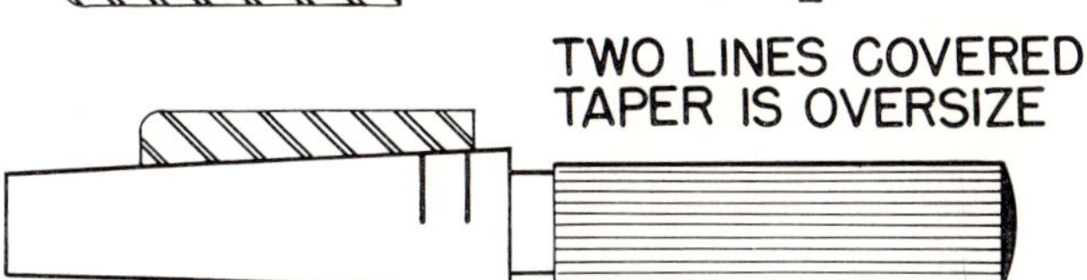

Figure d

Some taper gages have two lines scribed on the gaging member indicating the tolerance. When the gage is inserted to fit the taper snugly, one of the two lines should not be visible (see Figure *b*). If the gage fits snugly to either line, the taper is considered within the limits. If the tapered hole that is being checked is undersize, the gage will show both lines plainly (see Figure *c*).
When the tapered hole is oversize both lines will disappear from sight (see Figure *d*).

LAST-WORD INDICATOR

When using a last-word indicator, the contact point B should always rest on the surface being checked as close to C as possible (see Figure *a*).

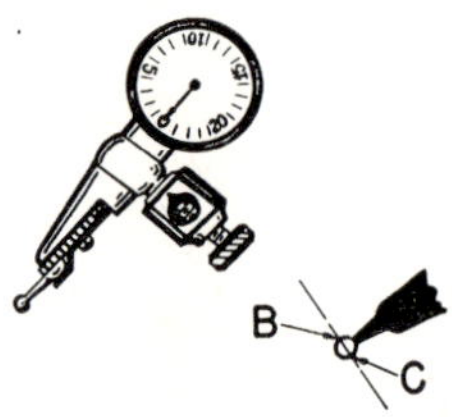

Figure a

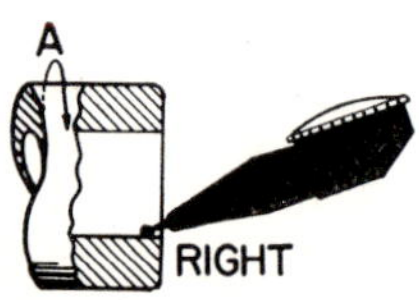

Figure b

Figure c

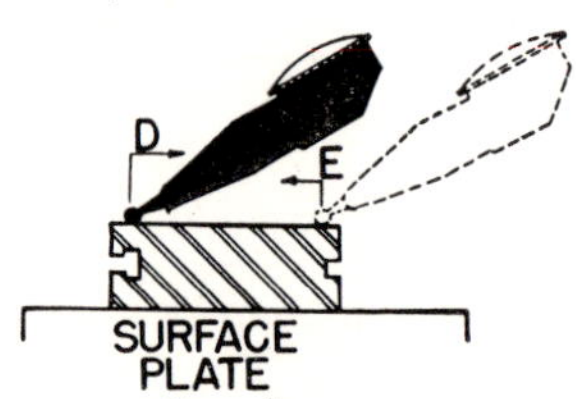

Figure d

Figure *b* shows the correct use of the indicator for checking the runout of a hole. The work should be rotated in direction A toward the inspector.

Figure *c* shows the wrong use of the indicator.

When inspecting a piece of work for flatness, the indicator should be moved in direction D and not in direction E (see Figure *d*).

CONCENTRICITY SETUP

This is a fast and accurate setup for checking the concentricity of an outside diameter from a hole.

Figure a

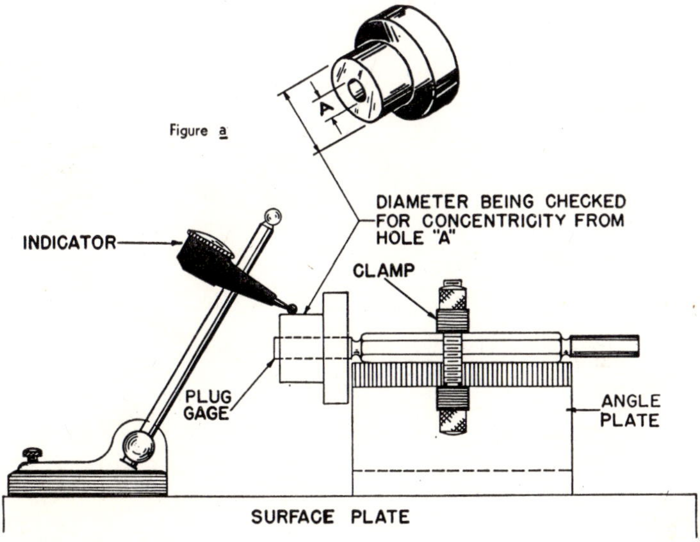

Many pieces can be inspected rapidly and accurately by putting the part on a plug gage, placing an indicator on the diameter to be checked, and rotating the work (see Figure *a*).

ANGLE-PLATE SETUP

Before setting up any work on an angle plate, this plate must be checked for its perpendicularity to the surface plate (see Figure a).

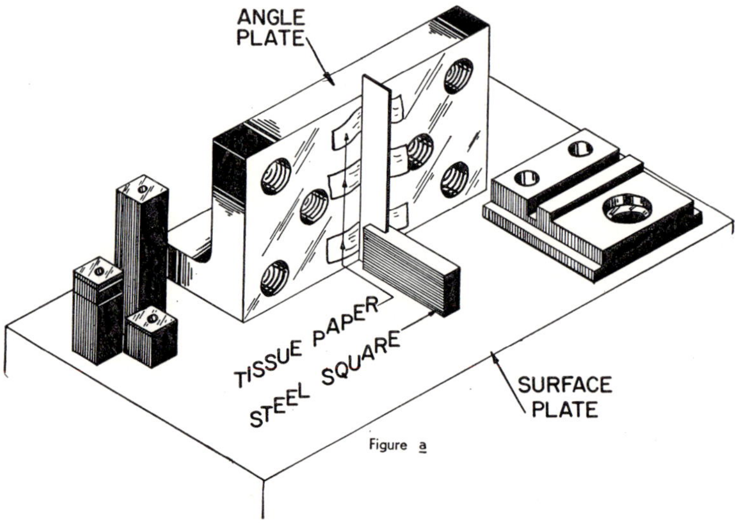

Figure a

1. Put a steel square against the face of the angle plate and place three pieces of tissue paper (of the same thickness) between the angle plate and the blade of the steel square.
2. Hold the steel square against the tissue paper and the angle plate. *Do not force!*
3. Try to remove all three pieces of tissue paper, one at a time. Pull on the tissue paper lightly and evenly. *Do not jerk!*
4. If the top piece of tissue can be pulled out and the bottom one holds, the angle plate is leaning toward the back. If the bottom piece of tissue is pulled out and the top piece holds, the angle plate is leaning forward. If all three pieces of tissue paper hold, the angle plate is perpendicular to the surface plate.

At no time should the middle piece of tissue paper be tight and the two end pieces loose; if anything, the top and bottom tissues should be tight and the middle tissue slightly loose.

SIZE-BLOCK SETUP

Size blocks are very useful for checking slots of various sizes and shapes.

Checking slots or grooves with size blocks is fast and accurate.

Several applications are show in Figure *a*.

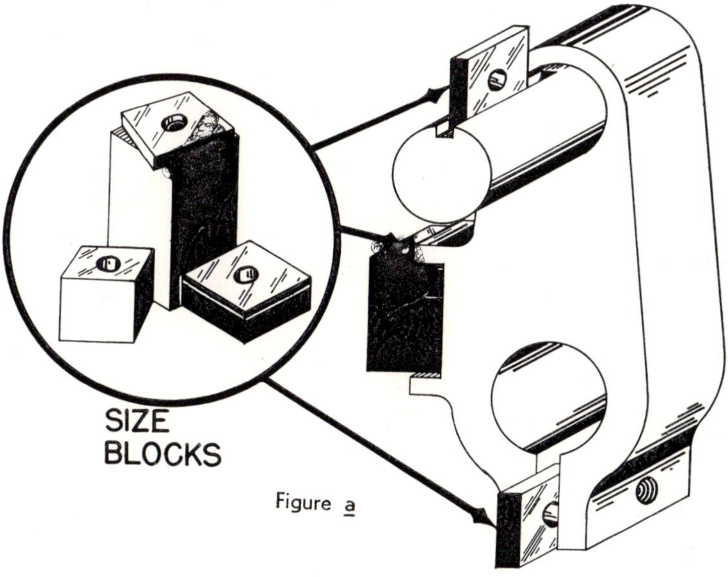

Figure a

COMPARATOR SETUP

One or more parts can be easily and rapidly checked for size by setting the indicator on a comparator with size blocks of the proper height.

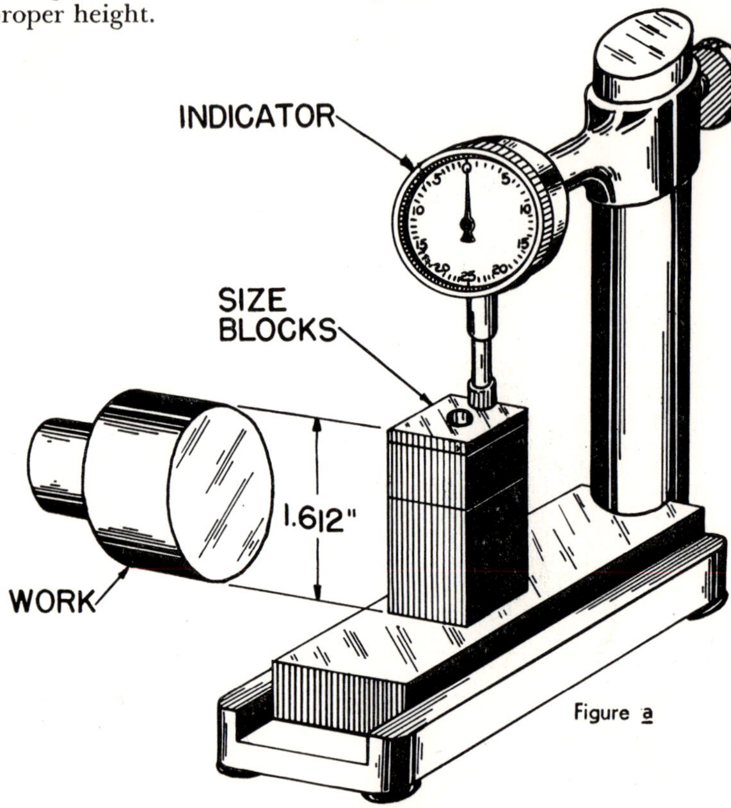

Figure *a*

After the indicator is set and firmly locked, the size blocks are removed and the part or parts that are to be checked can then be inspected by placing them one at a time under the indicator (see Figure *a*).

CHECKING TWO DIFFERENT DIAMETERS WITH ONE SETUP

If you have to check a part as illustrated in part drawing No. 1238, both diameters can be checked with a single setup as shown in Figures *a* and *b*.

1. Set an indicator to a 2-inch size block. In other words, set an indicator to the largest diameter (see part drawng No. 1238).

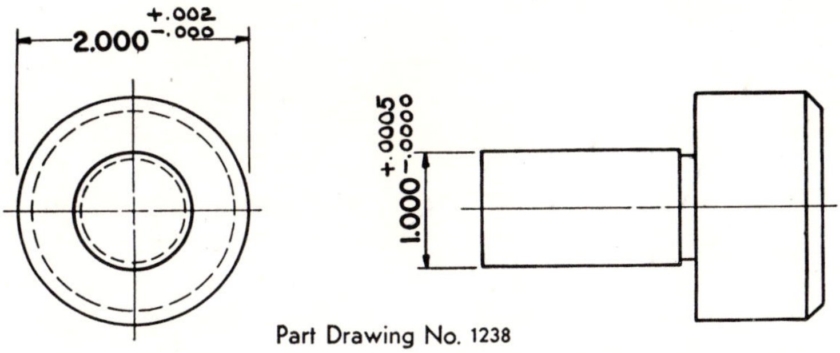

Part Drawing No. 1238

2. Pass this diameter between the surface plate and the indicator (see Figure *a*). The indicator hand must make the same number of revolutions and come to the same stop as it did when it was originally set to the size block.

Without upsetting the indicator, the size of the smaller diameter of the same part can also be checked as follows:

a. Subtract half of the small diameter from half of the large diameter.

b. Multiply the result by 2. The answer will be the height of the size block that has to be put under the small diameter to bring it up to the proper height (see Figure *b*).

c. Pass that diameter with the size block between the surface plate and the indicator again. The indicator hand must make the same number of revolutions and come to the same stop as it did when the larger diameter was checked.

Example:

Half of the large diameter	1.000 inch
Half of the small diameter	0.500 inch
	0.500 inch
	$\times 2$
Height of the size block of Figure *b* =	1.000 inch

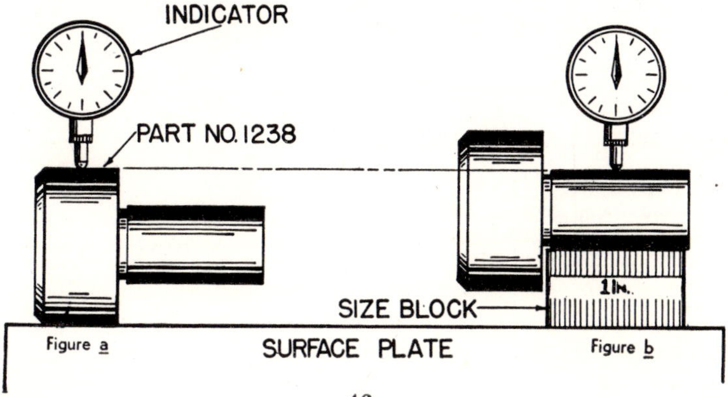

Figure a Figure b

VEE-BLOCK AND SIZE-BLOCK SETUP

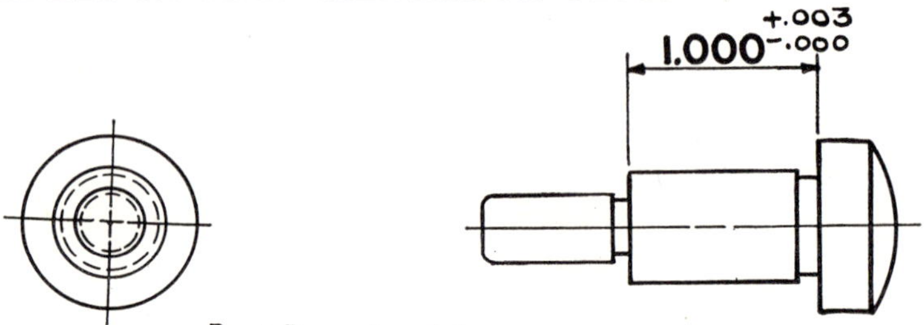

Part Drawing No. 1237

Another very effective but simple setup is made with vee blocks and size blocks. This setup can be used to check parts of many different shapes.

Figure *a* illustrates how to check the 1.000 $^{+0.003}_{-0.000}$ inch dimension of part No. 1237.

1. Set a vee block on end on the top of a surface plate.
2. Put a 1.000 inch size block, with a caliper jaw attached to it, on the top of the vee block.
3. Set the part in the vee block, with one shoulder resting on the same surface as the size block.
4. Reverse the switch lever on the indicator, so that the contact point will take a reading from under the head of the piece.
5. Set the indicator to the caliper jaw.
6. Move the indicator from the caliper jaw to the part being checked. If the reading from the caliper jaw to the part remains the same, the distance between the shoulder and the head of the part is just 1.000 inch.

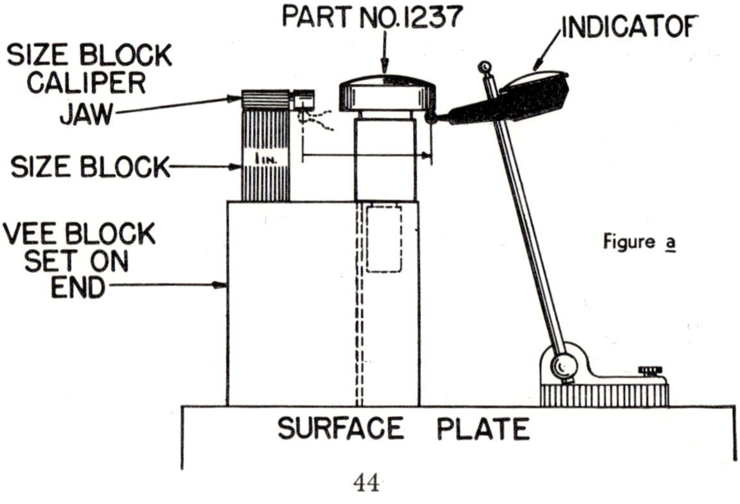

Figure a

CHECKING A 45° ANGLE

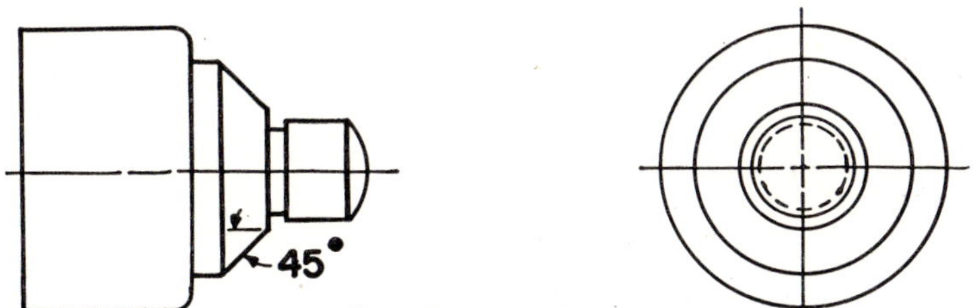

Part Drawing No. 1235

When a part, such as shown on part drawing No. 1235, is machined with a 45° angle and the position of the angle is such that it cannot be checked with a universal bevel protractor, set the part on a vee block as shown in Figure *a*.

Next set an indicator at one end of the angle surface and move it over to the opposite side of the angle to B. If the angle is 45°, the indicator will remain at its original setting.

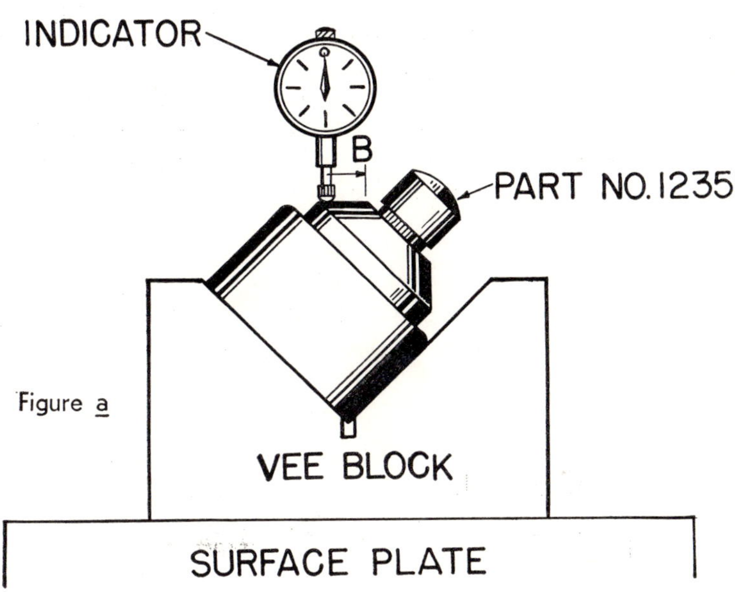

Figure a

SINE BAR

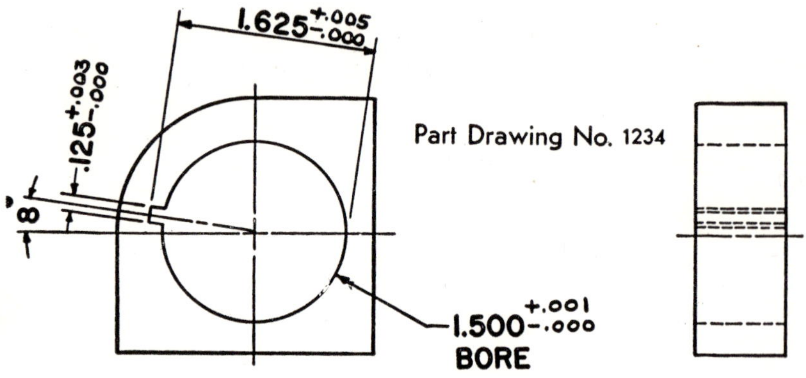

For checking the 8° angle of the $0.125 \begin{smallmatrix}+0.003\\-0.000\end{smallmatrix}$ inch key slot (see part drawing No. 1234) with a sine bar, use the following procedure:

1. Fit a size block to the 0.125 inch slot.
2. Set part No. 1234 on a sine bar, as shown in Figure *a*.
3. Referring to the table of constants for a 5-inch sine bar on page 139, you will find that the constant of 8° is 0.69585.
4. Build up size blocks to 0.69585 inches and place them under one cylinder, as in Figure *a*.
5. Set an indicator at point A on the size block and move it to point B. If the key slot is in the correct position, the indicator hand will not move when the indicator is moved from A to B on the size block.

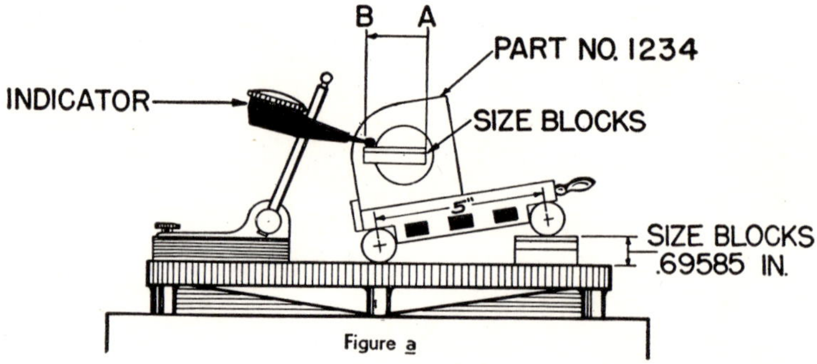

Figure a

A HOLE DRILLED AT RIGHT ANGLES TO A STEP

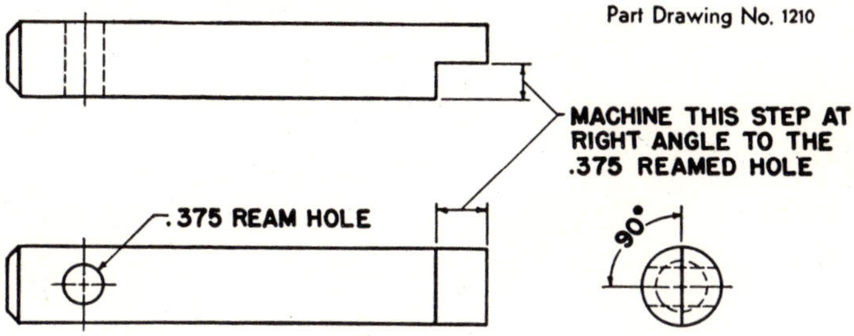

Part drawing No. 1210 shows a 0.375-inch hole drilled at right angles to a step.

A simple way of checking this part to see whether or not the hole is in the correct position is illustrated in Figure *a*.

1. Put a 0.375-inch drill blank or plug gage in the hole.
2. Set the piece in a vee block.
3. Put a square against the flat part of the step.
4. Set an indicator to zero over the drill blank or plug gage at A.
5. Move the indicator over to side B and pass it over the drill blank or plug gage at this point. If the indicator reads the same at B as it did at A, this means that the 0.375-inch hole is at right angles to the step.

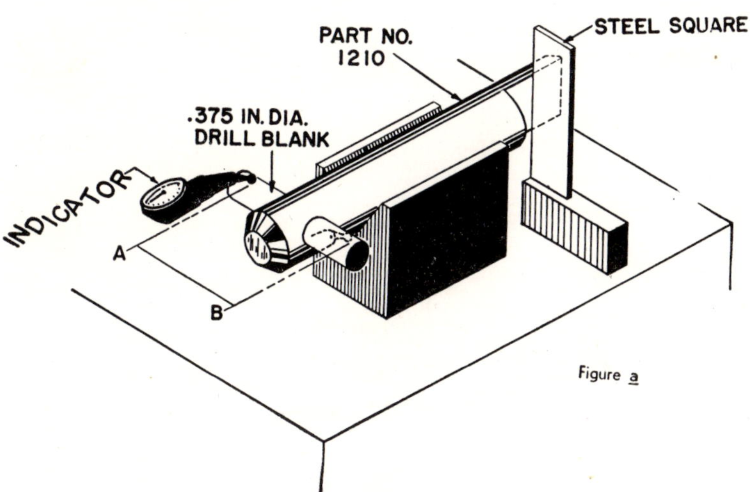

Figure a

BENCH-CENTER SETUP

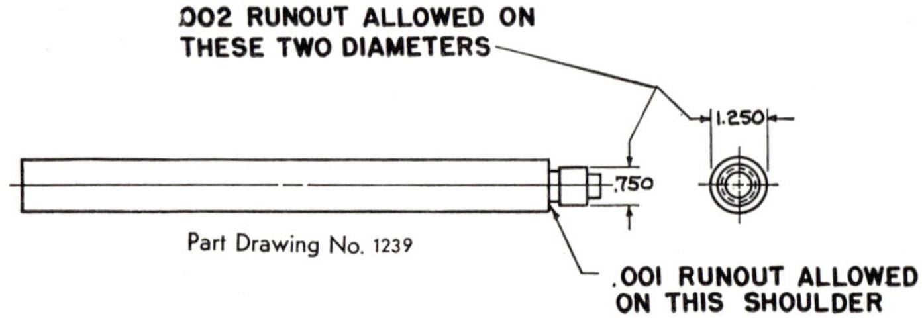

All runout and diameter sizes shown on part drawing No. 1239 can be checked with one setup as illustrated in Figure *a*.

1. Add B to A. 0.375 + 4.250 = 4.625 inches.
2. Set a dial indicator to size blocks that are built up to 4.625 inches and note where the revolution counter stops.
3. Add C to A. 0.625 + 4.250 = 4.875 inches.
4. Do not touch the original setting of the dial indicator, but pass a set of size blocks that are built up to 4.875 inches under the same indicator. Note where the big pointer and the revolution counter stop. Now the indicator is set to check the 0.750 and the 1.250 inch diameters.
5. By passing the dial indicator over the high point of one diameter at a time, the exact diameters will be found. If the indicator pointers repeat their original setting, the diameters are machined to the drawing.
6. Check the runout of these two diameters by leaving the indicator on the diameter being checked and rotating the work in direction D.
7. Place another indicator against the shoulder of the work and again rotate the work in direction D to measure the runout of the shoulder.

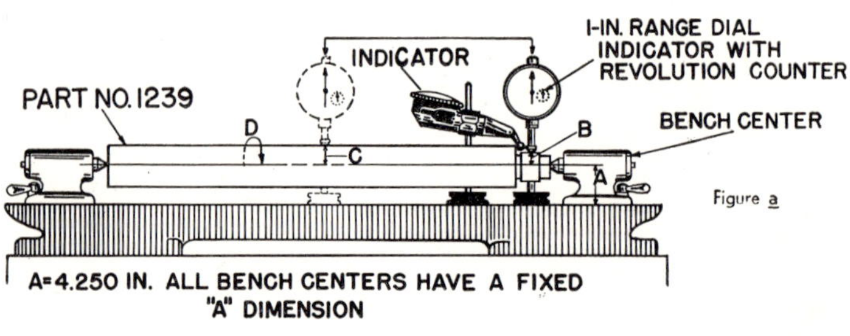

Figure *a*

SHOULDER RUNOUT

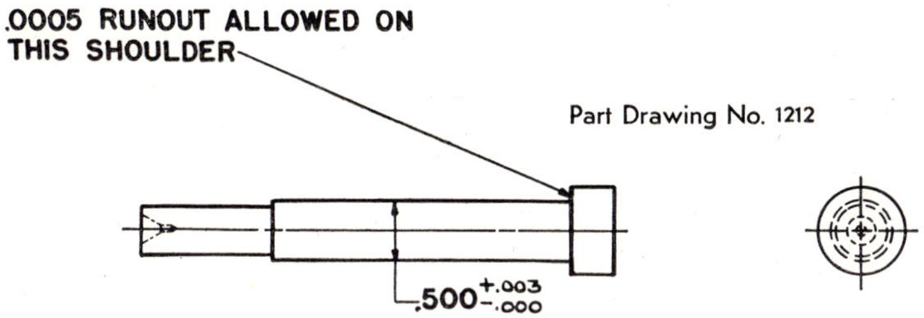

To check the 0.0005-inch shoulder runout from the $0.500 \begin{smallmatrix} +0.003 \\ -0.000 \end{smallmatrix}$ inch outside diameter of the piece shown on part drawing No. 1211, proceed as illustrated in Figure *a:*

1. Set the part in a vee block.
2. Put a steel ball between the part and the angle plate.
3. Set an indicator against the shoulder to be checked.
4. Hold the part (lightly) against the steel ball and the angle plate to prevent the steel ball from falling out.
5. Rotate the shoulder of the part against the contact point of the indicator and read the total runout.

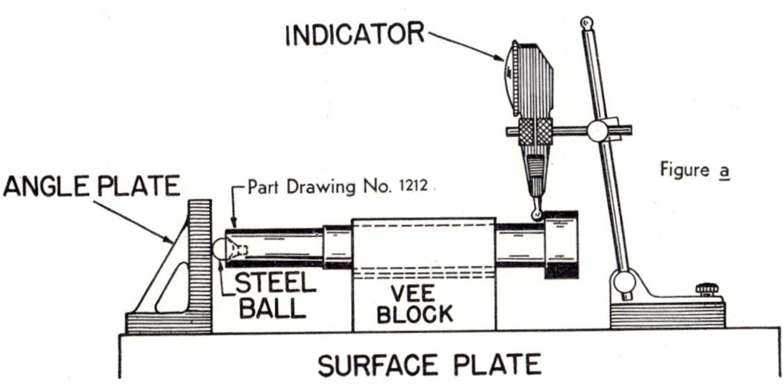

Figure a

CHECKING THE CENTER DISTANCE BETEWEEN TWO HOLES

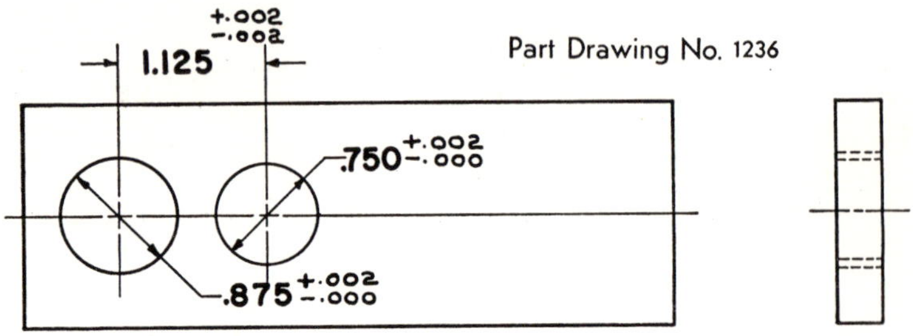

Part Drawing No. 1236

The 1.125-inch center distance between the $0.875 \genfrac{}{}{0pt}{}{+\,0.002}{-\,0.000}$ inch and $0.750 \genfrac{}{}{0pt}{}{+\,0.002}{-\,0.000}$ inch holes shown on part drawing No. 1236 can be checked accurately as illustrated in Figure *a*.

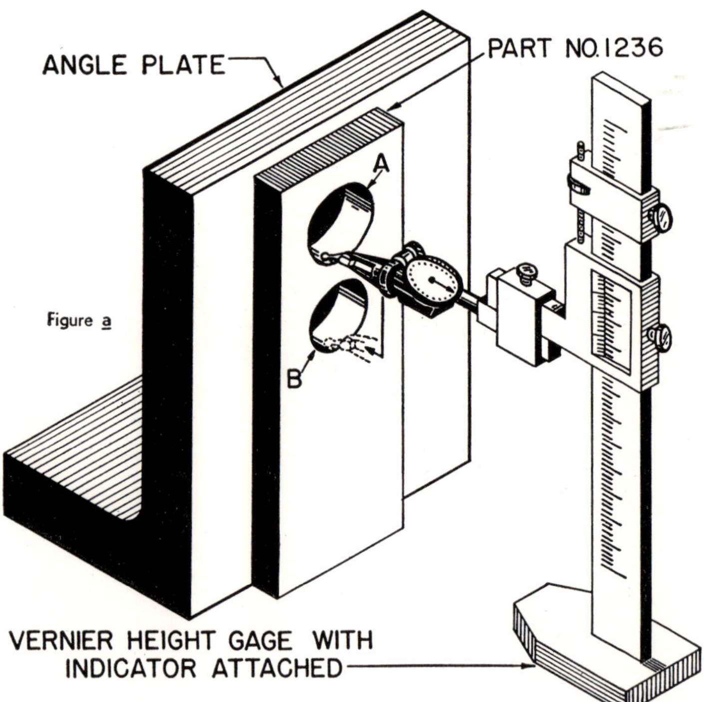

Figure a

1. Set the part on a surface plate and against an angle plate.

2. Set the contact point of an indicator, assembled with a vernier height gage, to the lowest point of the 0.875-inch hole A.

3. Without moving the indicator from this position, turn the dial on the indicator so that the hand is on zero.

4. Read the vernier. Let us say that the reading is 4.750 inches.

5. Move the vernier slide down to the lowest point of the 0.750-inch hole B. *Do not touch the indicator!* Move the vernier slide until the indicator hand stops on zero again.

6. Read the vernier again at this point. Let us say that the reading is 3.6875 inches.

7. Subtract the 3.6875 inches from the 4.750-inch reading and from the result subtract half of the 0.750-inch hole. Add half of the 0.875-inch hole to the resulting figure. This will give 1.125 inches which is the center distance between the two holes.

Example:	4.7500	First reading
	—3.6875	second reading
	1.0625	
	—0.3750	half of the 0.750-inch hole
	0.6875	
	+0.4375	half of the 0.875-inch hole
	1.1250	the center distance between the two holes

CHECKING THE CENTER DISTANCE BETWEEN TWO SPHERICAL DEPRESSIONS

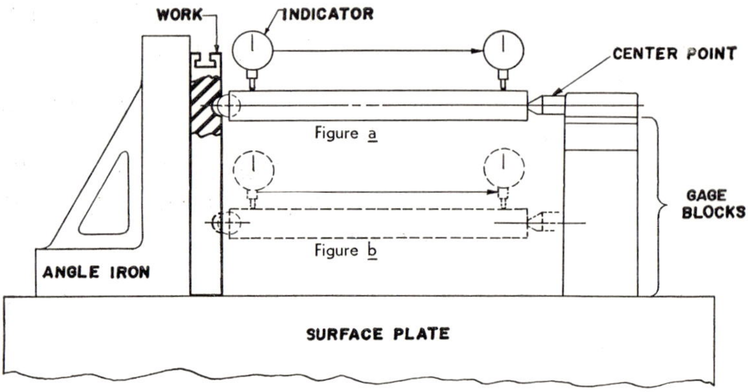

It is rather difficult to check the center distance between two spherical depressions, but it can be done by doing as suggested.

Mount the work piece to an angle plate. Place a steel ball that is slightly larger into the spherical depression. Back the steel ball up with a perfectly straight piece of drill rod that has a center hole in each end. Build up a set of gage blocks and center point attachment, so that the rod will indicate zero at both ends. Take the reading of this height from the vernier to which the indicator is attached (see Figure *a*).

Remove the ball and rod and place them into the bottom spherical depression (see Figure *b*). Build the gage blocks up so that the drill rod will indicate zero again at both ends. Take the reading of this height from the vernier and subtract it from the reading obtained in Figure *a*.

The result will be the distance between the centers of both spherical depressions.

CHECKING A TAPER HOLE WITH A TAPER-PLUG GAGE

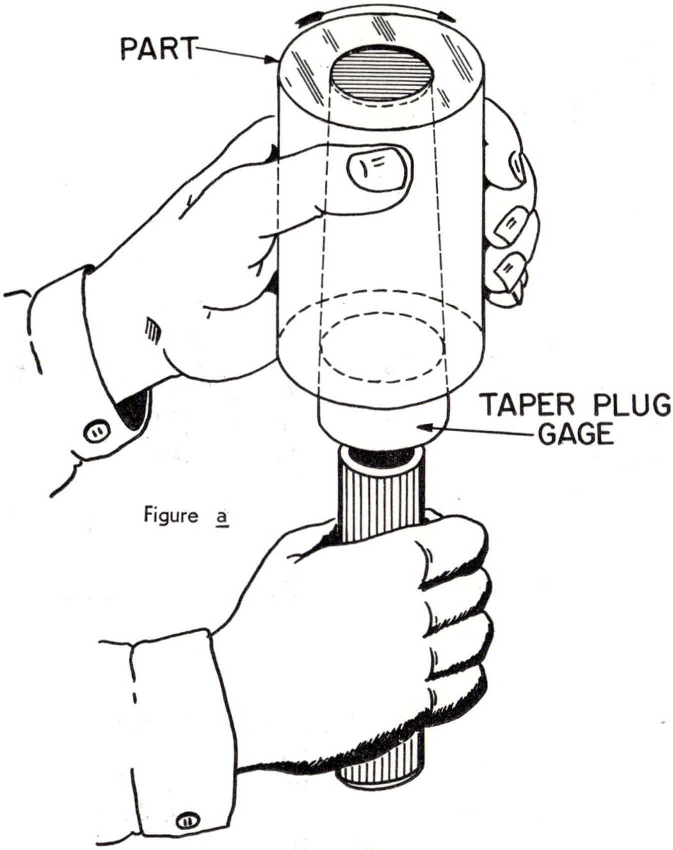

Figure *a*

To check the bearing of a taper hole and to get a true reading:

1. Put a light coat of Prussian blue on the gaging member of the taper gage.
2. Insert the taper gage into the taper hole of the part being checked with caution, so that the sides of the gage do not rub the side of the hole before the gage is completely in the hole.
3. Hold the part and the gage vertically.
4. Hold the gage firmly (do not let it turn).
5. Rotate the part in the direction of the arrow, pressing lightly downward with even pressure (see Figure *a*).
6. Remove the gage and the bearing of the hole will show up on the gaging member wherever the Prussian blue is removed

LAYING OUT A ROUGH CASTING

Laying out a rough casting may appear difficult to some inspectors, but it can be readily done by carrying out the following steps (see Figures *a* and *b*):

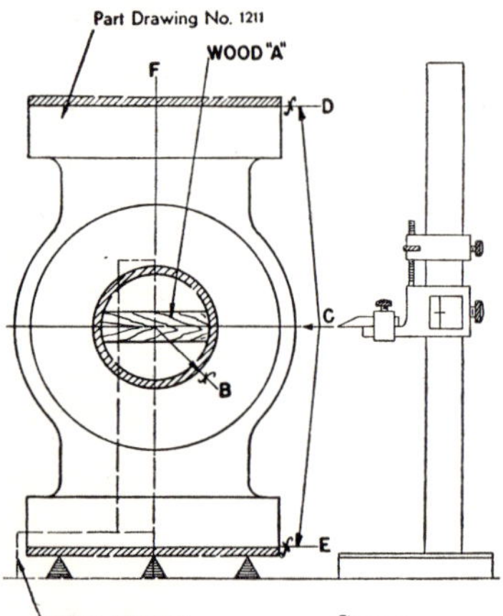

Figure a

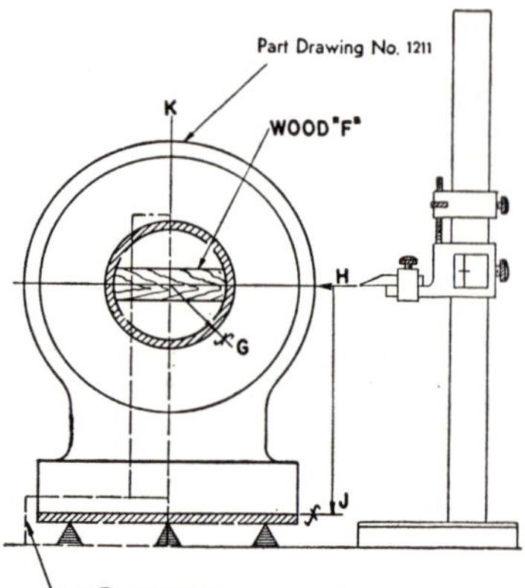

Figure b

1. Insert a piece of wood "A" in each of the three $2^{13}/_{16}$-inch holes (see part drawing No. 1211).
2. Paint the face of these holes and the outside diameter of the flanges with a layout dye or chalk. This will make the lines that will be scribed on the part later show up clearly.
3. Set the casting on three adjustable jacks (see Figure *a*) and place them about 120° apart.
4. Resting a steel square on the surface plate and the blade of the square against the face of the flange, adjust the jacks until the face of the flange is parallel to the blade of the square.
5. Mark the center lines C and F of the $2^{13}/_{16}$-inch hole on the piece of wood "A."
6. Set a pair of dividers to the radius of the finished hole B which is $1\frac{1}{2}$ inches.
7. Place the point of one of the caliper legs on the center line that was scribed on the piece of wood "A" and scribe the circle B of Figure *a*.

8. Set the point of a scriber connected with a vernier height gage, to the center line C.
9. Move the vernier slide up 6 inches from the setting and scribe a line around the outside diameter on the flange D.
10. From this setting, move the vernier slide down 12 inches to the bottom flange and scribe line E.
11. With a steel scale, measure the distance between the scribed lines and the nearest edge of the rough casting. The distance thus obtained is the amount of material that is left to be machined.
12. Remove the casting from the jacks and then place it back on the jacks again with hole B facing down (see Figure *b*). Square the casting in the same manner as described in Figure *a*, mark the center lines K and H, radius G, and flange thickness J in the same manner as described for Figure *a*.

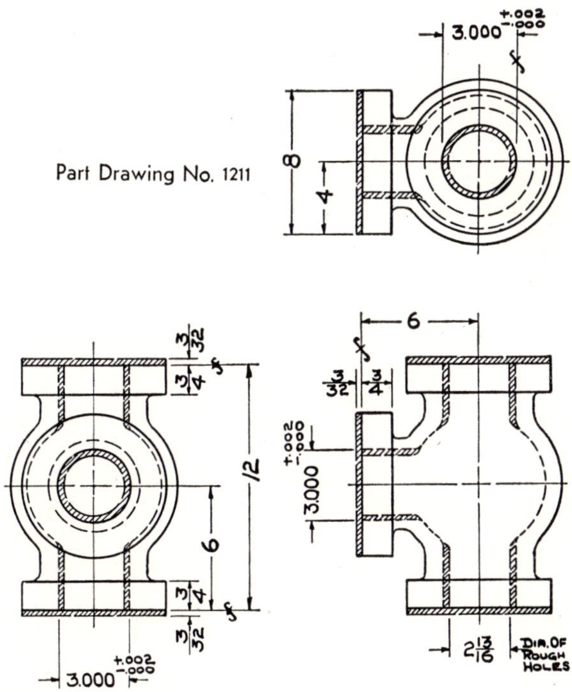

Part Drawing No. 1211

NOTE: PATTERN MAKER- ALLOW 3/32" FOR FINISH MACHINING.

SETTING UP LIGHT CASTINGS

When adjusting screw jacks that are supporting a light casting, they will invariably slip on the surface plate.

Slipping of these jacks can be eliminated by placing them on strips of tape as shown in Figure *a*.

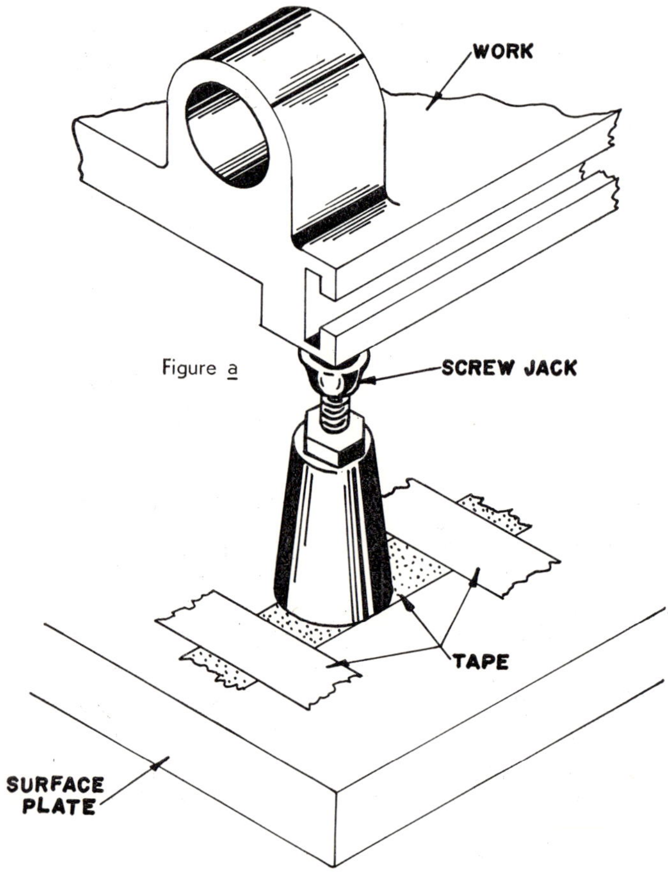

Figure a

SURFACE GEOMETRY

Imperfections on a surface finish belong to two categories.

The first is a roughness which consists of closely spaced irregularities such as successive passes of the metal cutting tool. The second is by the cutting action of grinding or polishing abrasives.

Figure *a*, shows surface waviness.

Figure a

Figure *b*, shows a surface having roughness and waviness.

Figure b

Figure *c*, shows surface roughness.

Figure c

OPTICAL FLATS

Concave surfaces are identified by a concentric-ring pattern (see Figure *b*). By pressing down on the center of the optical flat, the rings are brought toward the center (see Figure *a*).

Convex surface will remain unchanged when the flat is pressed in center, but if the flat is rocked, the pattern rings will shift (see Figure *c*).

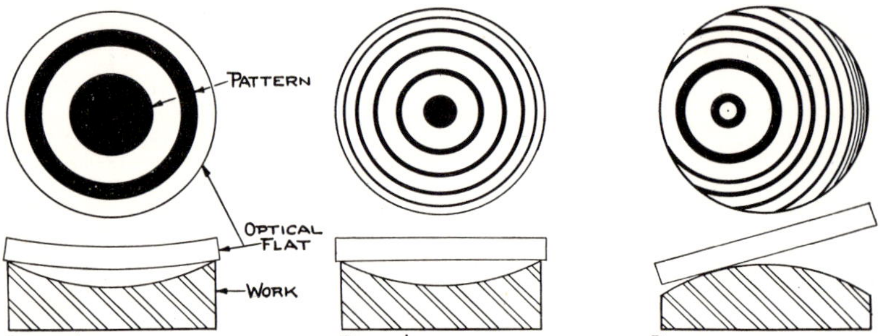

Figure a Figure b Figure c

A perfectly flat piece of work placed between two optical flats will show straight parallel lines or fringes. If the part is not flat, the lines will be uneven.

The optical flat is one of the basic instruments.

An optical flat is a disc made of high-quality quartz glass that has been ground as near to perfect flatness and parallelism as it is possible.

Flats depend on the action of light-wave interference for their value.

The elements used to make a simple check are a light source, an optical flat, a small part to be checked, and a flat surface on which the part can be set.

If the part and the flat are clean and the inspector views the reflected light from a point above almost in line with the light, a group of parallel shadow lines will appear when the optical flat is pressed down.

Shadow lines are also referred to as fringes.

A thin wedge of air is actually present between the bottom surface of the optical flat and the surface of the part being checked.

SURFACE SHAPES

The most common surface shapes are flat, concave, convex and irregular.

To acquaint the inspector with their appearance, Figures *a, b, c* and *d* clearly illustrates them.

FLAT

Figure a

CONCAVE

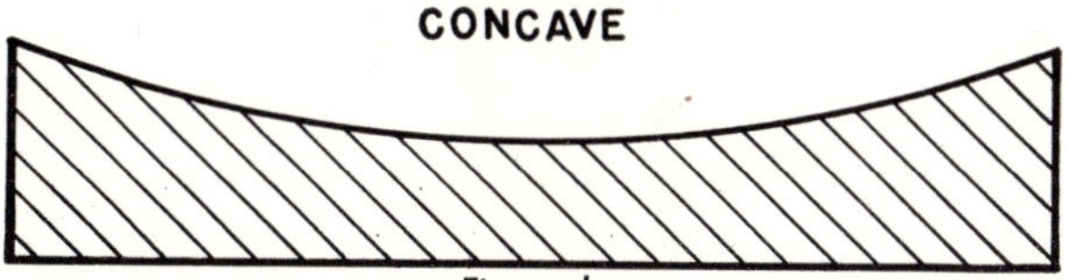

Figure b

CONVEX

Figure c

IRREGULAR

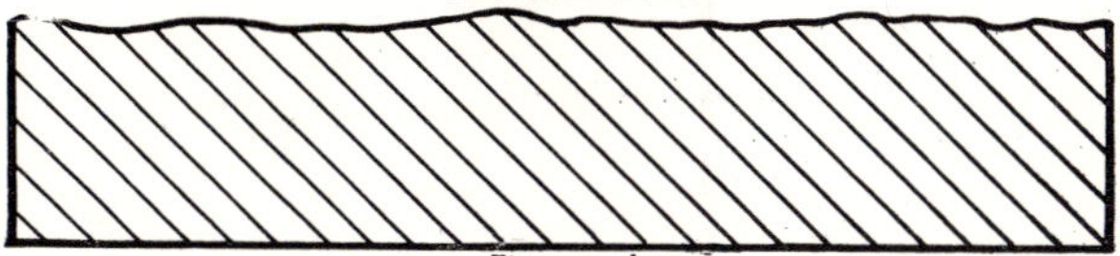

Figure d

"PILOT" CYLINDRICAL PLUG GAGE

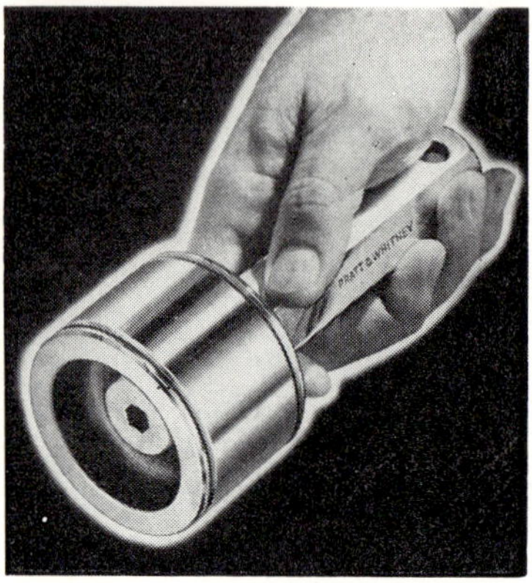

Figure *a*. This gage can be presented at any angle and as it is straightened up, it practically falls into the hole without careful guidance or pressure.

Figures *b* to *d* illustrate the correct use of this type of plug gage.

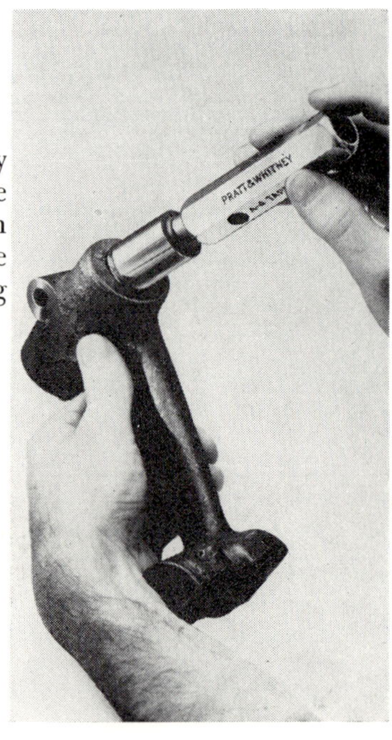

Figure *b*. The gage is only 0.0001 inch smaller than the hole and the piece is being held in the hand which would make gage entry with an ordinary plug difficult.

Figure *c*. Notice how easily the "Pilot" enters the hole without pressure and without jamming.

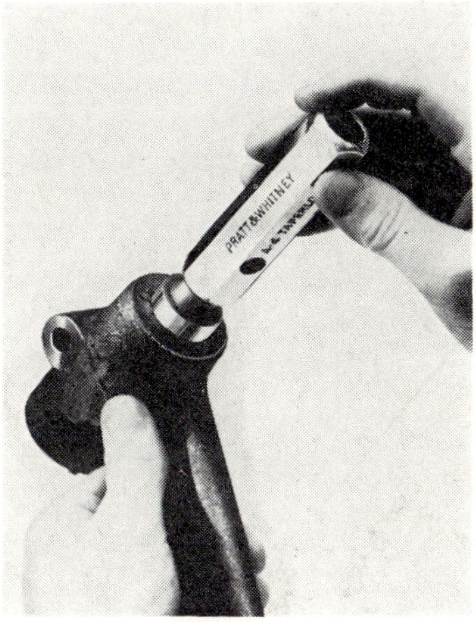

Figure *d*. The hole is only 0.0001 inch larger than the gage. Yet the "Pilot" can be presented at any angle and it falls in almost automatically.

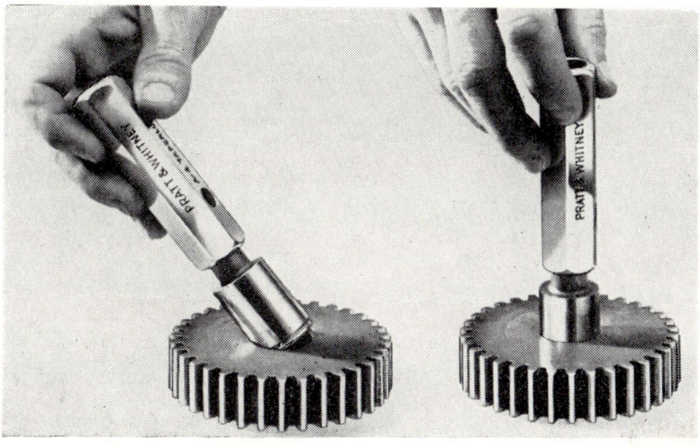

J-S ROLL THREAD SNAP GAGES

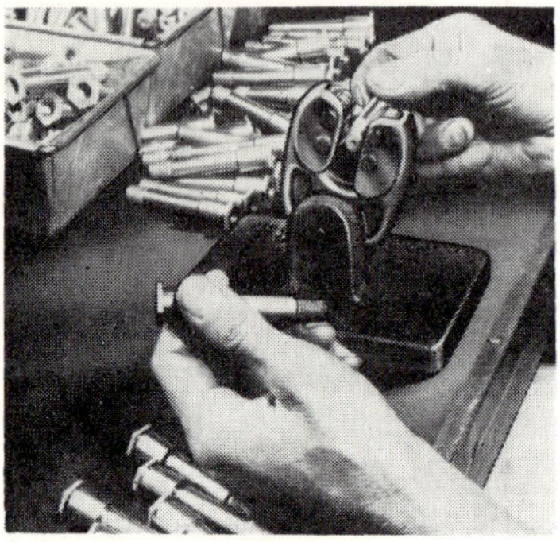

Figure *a*

These are go and not-go limit gages for the inspection of external threads. The gaging rolls have annular rotating ribs for engaging the threads and are mounted on eccentric pins supported in a U-shaped frame for adjustment (see Figure *a* and *b*).

J-S roll thread snap gages incorporate the following desirable features required for checking male threads:

1. Go and not-go gaging elements are combined in one page for checking minimum and maximum limits in one pass, rapidly.
2. The rolls rotate on contact which increases the gage life and reduces the hazard of checking parts with worn gages.
3. Inspection is analytical, revealing errors in lead, profile, angle, roundness, straightness, and diameter as well as detecting oversized or undersized threads.
4. Right- and left-hand threads of the same nominal size can be checked with the same gage.

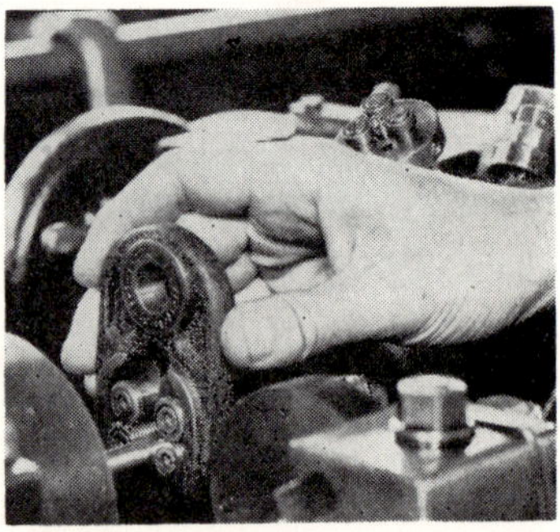

Figure *b*

ADJUSTABLE LIMIT SNAP GAGES

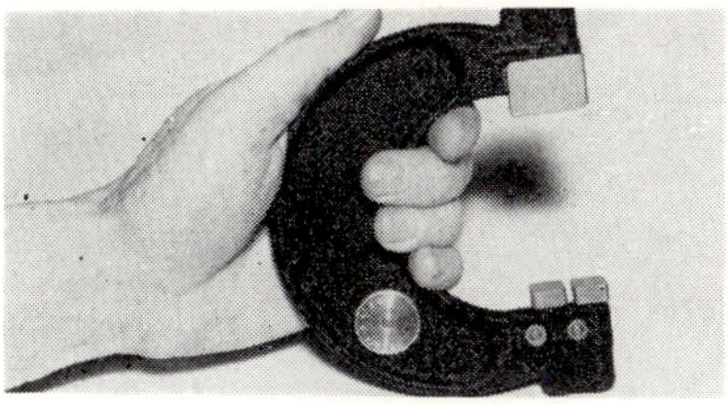

Figure *a*

Figure *b*

This gage is an ideal combination of light weight, great strength, and rigidity with ease of anvil adjustment, positive anvil locking, and high accuracy (see Figure *a*).

Figure *c*

It is an ideal gage for the limit system wherever diameters or lengths are to be inspected.

Figures *b, c,* and *d* show the gage in use.

Figure *e* illustrates how the limit snap gage can be set with gage-blocks precision.

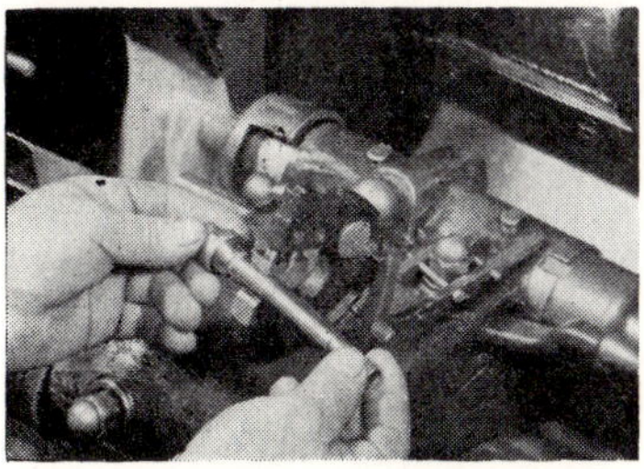

Figure *d*

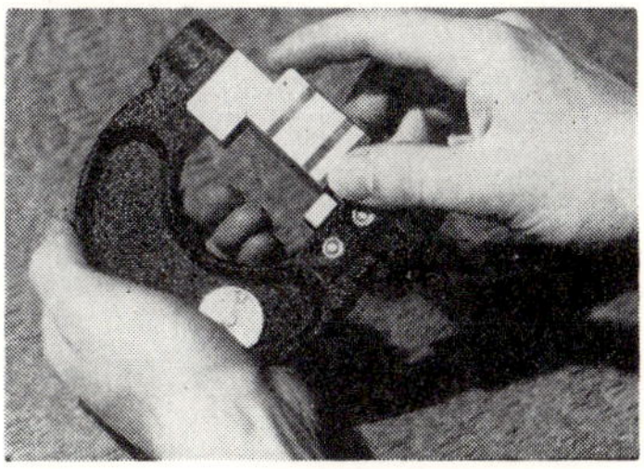

Figure *e*

RING AND PLUG THREAD GAGES

The ability of threaded parts to assemble readily and accurately depends on the precision with which the threads are made.

By checking both the external and internal threads of each regulator part, inspectors can accept only those whose thread conforms to the specified standard.

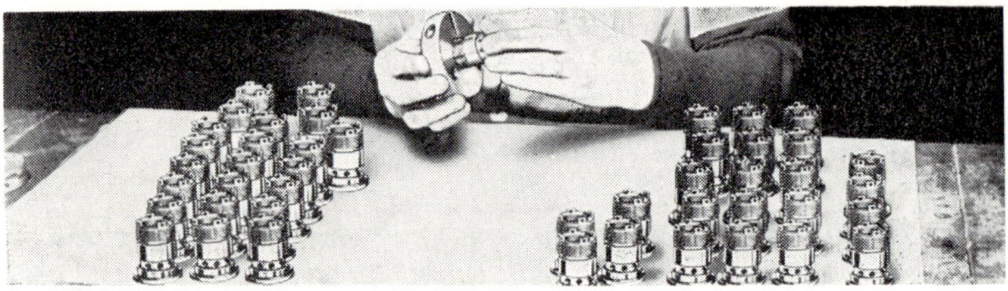

Figure *a*

Figure *a* shows the checking of an external thread with a thread ring gage.

Figure *b*

Figure *b* shows the checking of an internal thread with a thread plug gage.

HOKE PRECISION GAGE BLOCKS

Hoke precision gage blocks are special alloy-steel blocks, approximately 0.950 inch square, heat treated to maximum hardness.

The square shape facilitates handling and provides large symmetrical gaging surfaces which are guaranteed to be flat, parallel, and accurate within a few millionths of an inch.

These blocks range from 0.010 inch up to 20 inches in size.

GAGE-BLOCK AND TIE-ROD ASSEMBLY

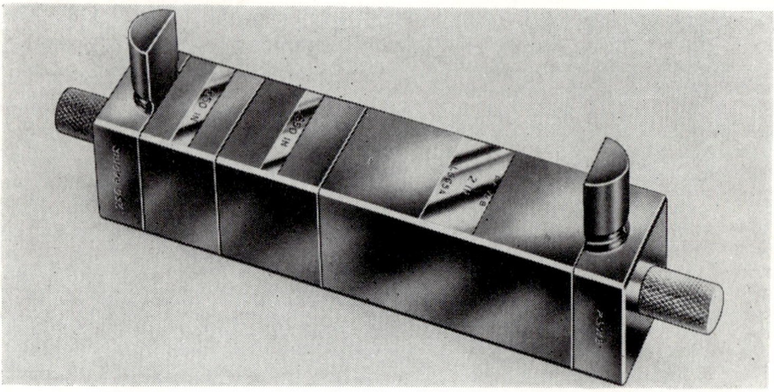

Figure *a*

Figure *a* illustrates how adjustable tie rods are used to make up a dimension for an internal precision measurement.

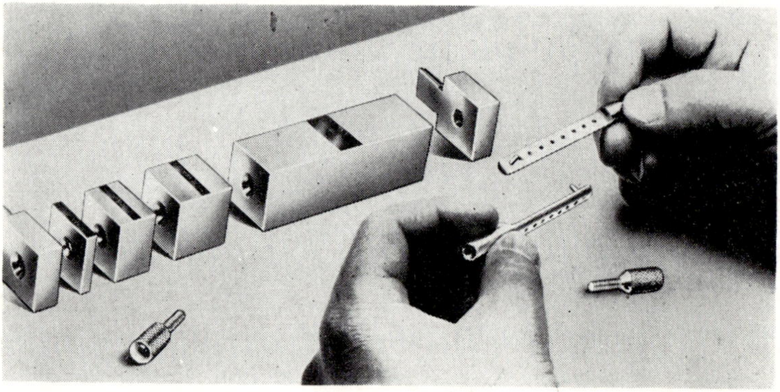

Figure *b*

Figure *b* shows the complete gage-block and tie-rod assembly.

PRECISION GAGE BLOCKS

Precision gage blocks should be in every inspection department. They can be used in stacks to make any desired dimensions with the combined error of all the blocks still negligible. They are used for everything requiring precision from checking a micrometer to inspecting complicated fixtures.

They are available in two styles. The USA blocks are rectangular (see Figure *a*). The Hoke blocks are square (see Figure *b*).

In wringing two blocks together, be careful not to damage them.

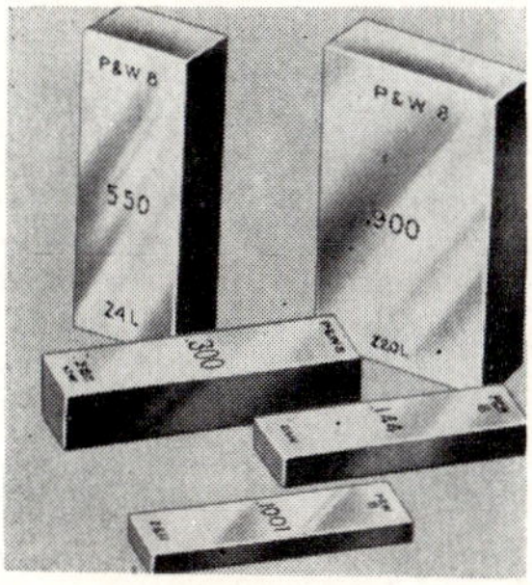

Figure *a*

Figure *b*

Figures *c* to *f* indicate the best method of joining two blocks.

Put the blocks together as shown in Figure *c*. This minimizes the danger of the corner of one block scratching the other.

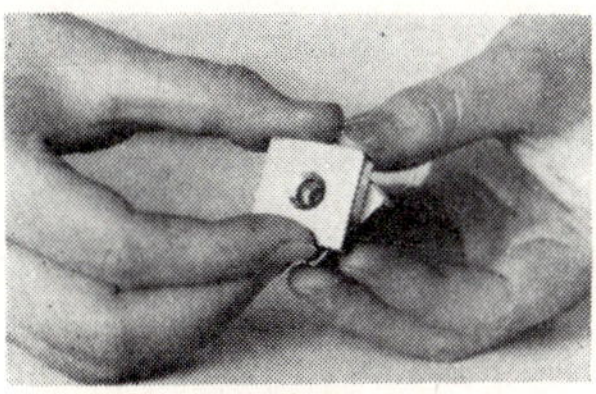

Figure *c*

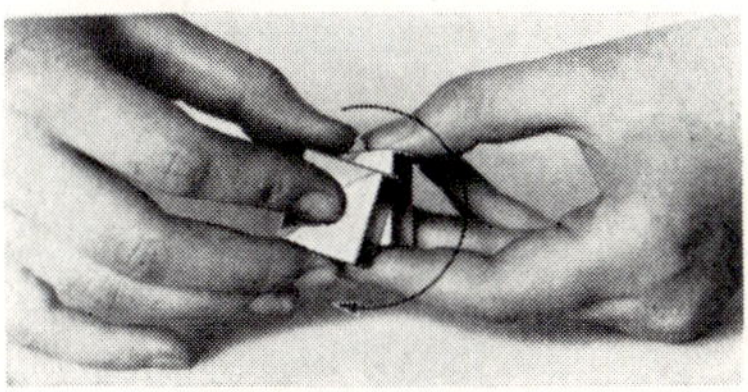

Figure *d*

Slide the upper block over the lower with a circular motion (see Figure *d*). If they are clean, they will begin to take hold. Foreign matter can be detected readily and should be removed before it damages either block.

Slide the upper block half out of engagement, using light pressure (see Figure *e*).

Slide the upper block back into full engagement under light pressure and wring them together ready for use (see Figure *f*). Worn or dirty blocks will not wring.

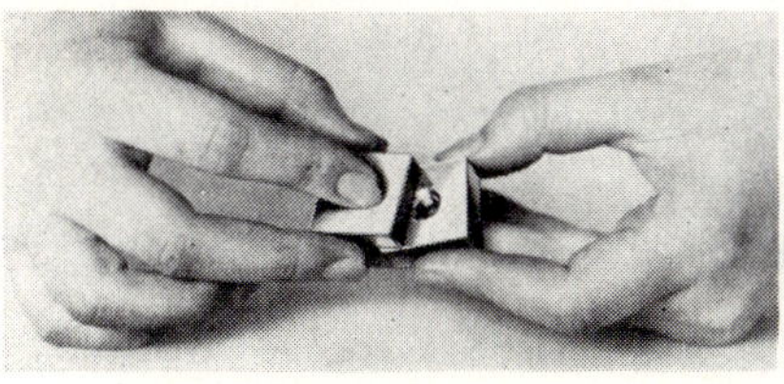

Figure *e*

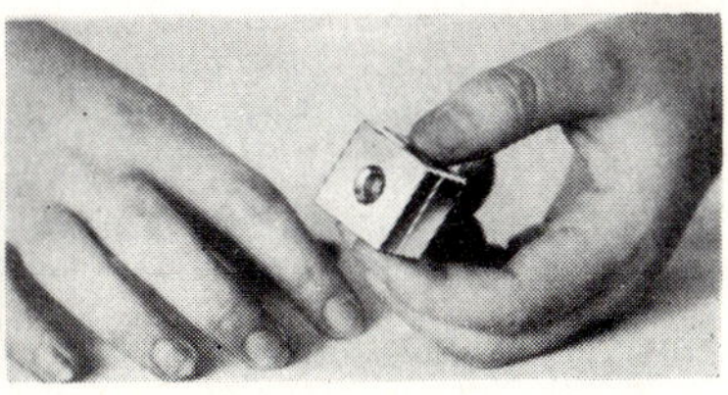

Figure *f*

ELECTRICAL HARNESS LACING

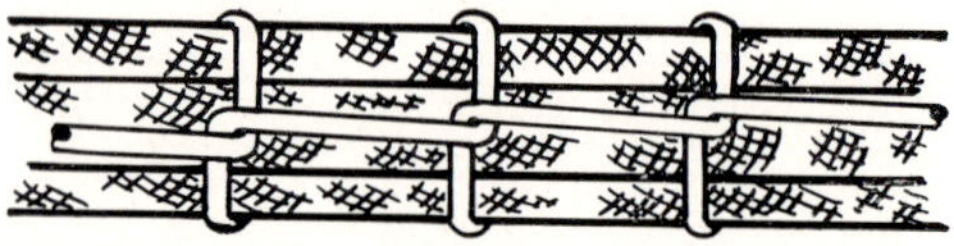

Figure a

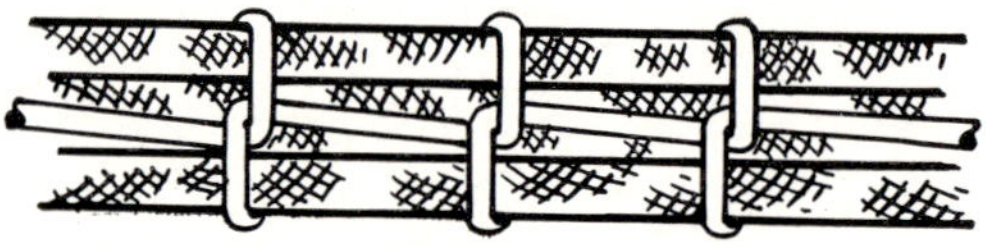

Figure b

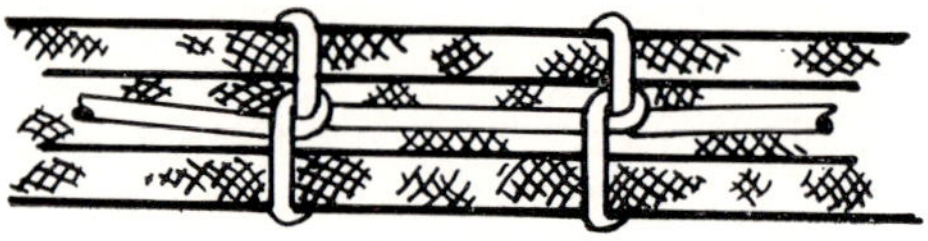

Figure c

There are three basic methods that are used to lace electrical harnesses.

Figures *a, b* and *c* clearly show them. The method shown in Figure *a,* called the half hitch, is not as practical as those methods that are show in Figures *b* and *c* because if the lacing should break at any point all the rest of the lacing would unwind allowing the coated wire of the harness to become loose.

The overhand knot shown in Figure *b,* or the figure of eight knot that is shown in Figure *c* are the more practical methods used because they are less apt to unravel if they should become broken or cut. Usually the method that is shown in Figure *b* is adequate for most requirements.

The distance between the knots can be from 1 to 1½ the diameter of the cable. However, no minimum space is required.

Plastic coated wires should not be laced too tight, if they are the cord is liable to cut through the insulation.

LOCKWIRING

Lockwiring is a positive method of securing in place bolts, screws, studs and nuts which hold together parts of aircraft engines that could become loose under vibration.

Two or more parts can be held together by this method, and if one of the parts should become loose it will have a tendency to tighten up on the lockwiring, therefore, preventing that part from becoming separated from its assembly.

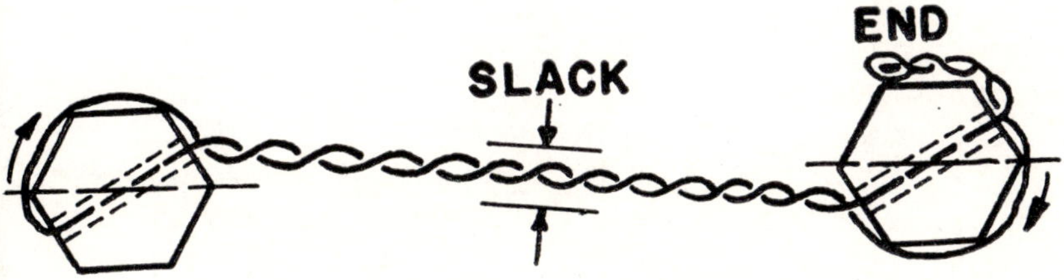

Figure a

Lockwiring must be tight but never overstressed. If it is too tight it might break under a jolt.

Approximately 0.125 inch maximum slack in either direction is permissible.

The end of the wire should be cut off between 1½ to 4 turns from the part and bent down. See Figure *a*.

THE THERMOMETER AND HOW IT WORKS

Place the thermometer into a container of melting ice. The lowest point which the liquid drops in the container will be the freezing point of water or the melting point of ice. On the Fahrenheit scale, this point is marked 32°F (see Figure *a*). On the Centigrade scale, this point is marked 0°C (see Figure *b*). Now if the thermometer is placed in boiling water the highest point reached by the liquid in the tube is called the boiling point of water. On the Fahrenheit scale it is marked 212°F (see Figure *a*). On the Centigrade scale, it is marked 100°C (see Figure *b*).

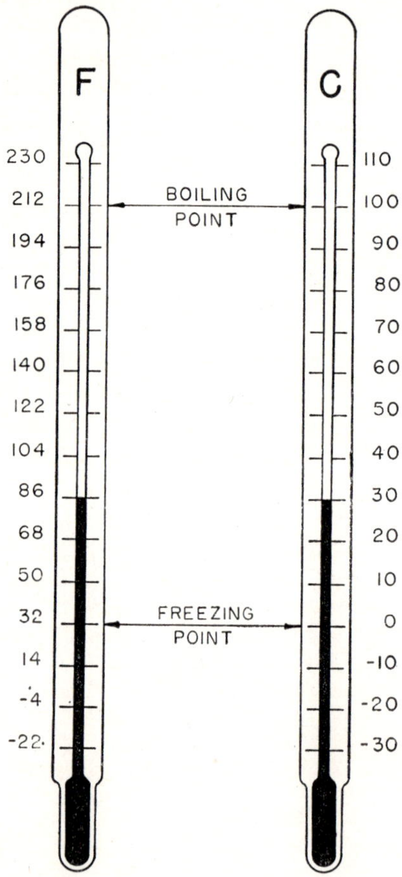

Figure a   Figure b

The space between the 32°F and 212°F is divided into 180 equal parts. On the Centigrade scale—as the name indicates—the space between 0°C and 100°C is divided into 100 equal parts.

Check a thermometer for the following:

1. Is it of the Centigrade or Fahrenheit type?
2. Is the glass chipped or cracked?
3. Are all of the numerals and graduations legible and in accordance with the drawing?

BASIC HELICAL SPRING DESIGNS

Helical springs are formed by winding wire of any shape into cylindrical coils. Heavy springs are formed while the steel is hot. Smaller springs are usually made of tempered steel wire, called music wire.

Figure *a*, shows an EXTENSION SPRING, Figure *b*, a COMPRESSION SPRING, Figure *c* a CORD SPRING, and Figure *d* a TORSION SPRING.

When inspecting helical springs visually, it is important that the following checks are made:

Figure a

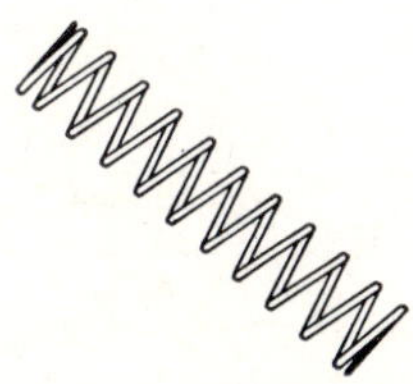

Figure b

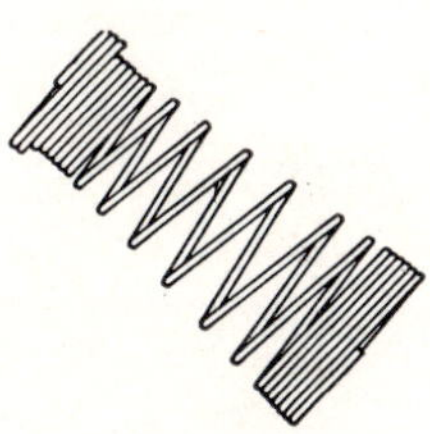

Figure c

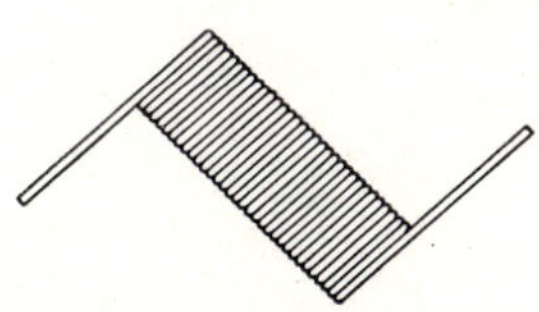

Figure d

1. Type called for
2. Active coils
3. Body length
4. Free length
5. Pitch
6. Total coils
7. Wire gage

FUNDAMENTAL FEATURES OF A THREAD

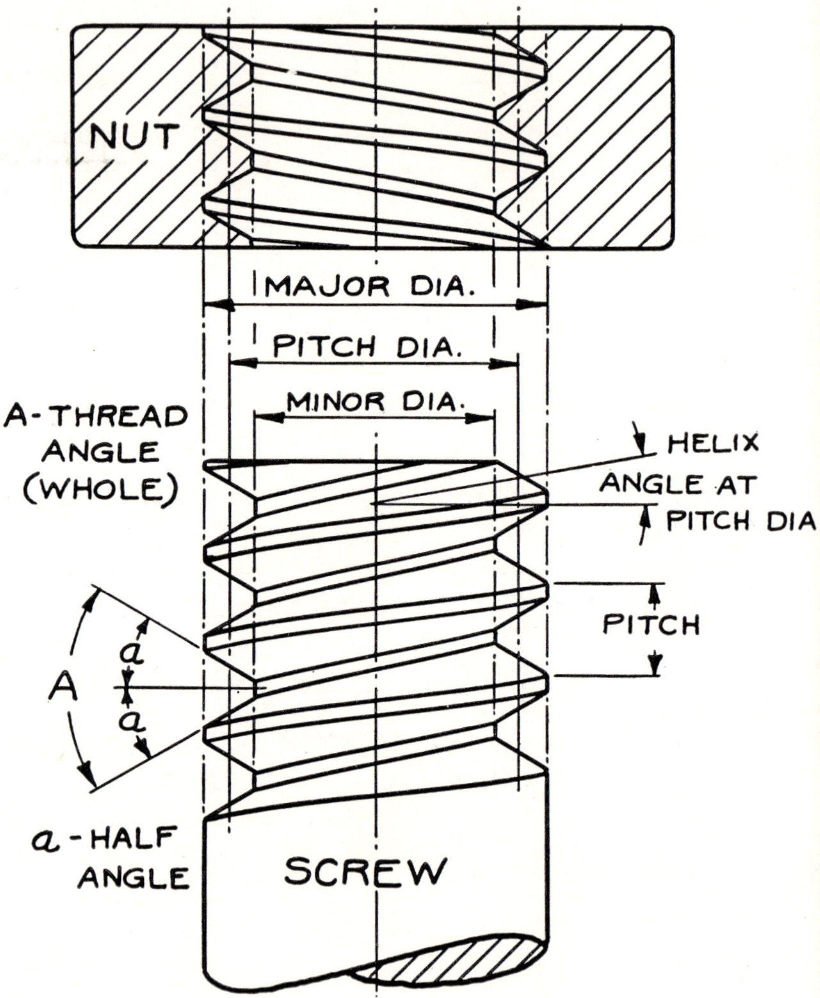

"BASIC" POINTS IN THREAD MEASURING

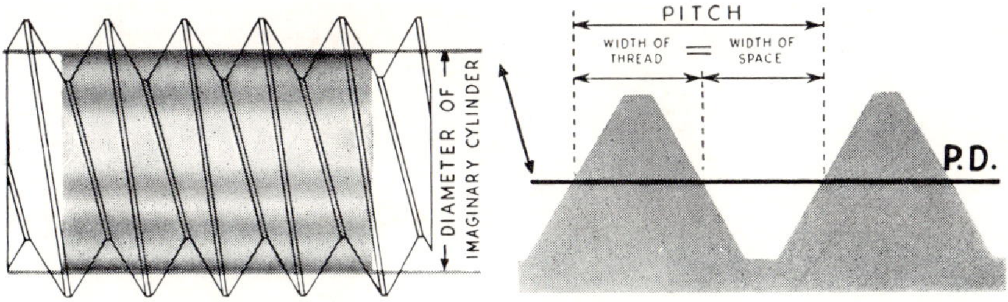

All measurements of a thread have a controlling point or base from which to start.

The control point in the case of a screw thread is called the *Basic* or theoretically correct size, which is calculated on the basis of a full form thread. On a given screw thread there is the *Basic Major Diameter* and *Basic Minor Diameter.* In practice, it is impossible to form screw threads to their exact theoretical or basic sizes, but it is possible to establish limits which the elevation should not exceed. They are called the *Minimum* and *Maximum* Limits. When the product is not larger than the *Maximum Limit* it is within the required size limits.

Basic size is not always between the *Minimum* and *Maximum Limits.* The tolerances for external threads will be below *Basic* and for internal threads, above *Basic.*

Actually the *Basic* dimensions follow the contours of the theoretically perfect thread.

On a straight screw thread, the outside diameter of an imaginary cylinder would pass through the threads at points that make the width of the threads and the width of the spaces cut by the outside diameter of the cylinder equal in size.

THE MICROINCH SCALE

The millionth or microinch is the basis for inspection of the United States standard inch.

A partial breakdown of a one-thousandths-inch scale is shown in Figure *a*.

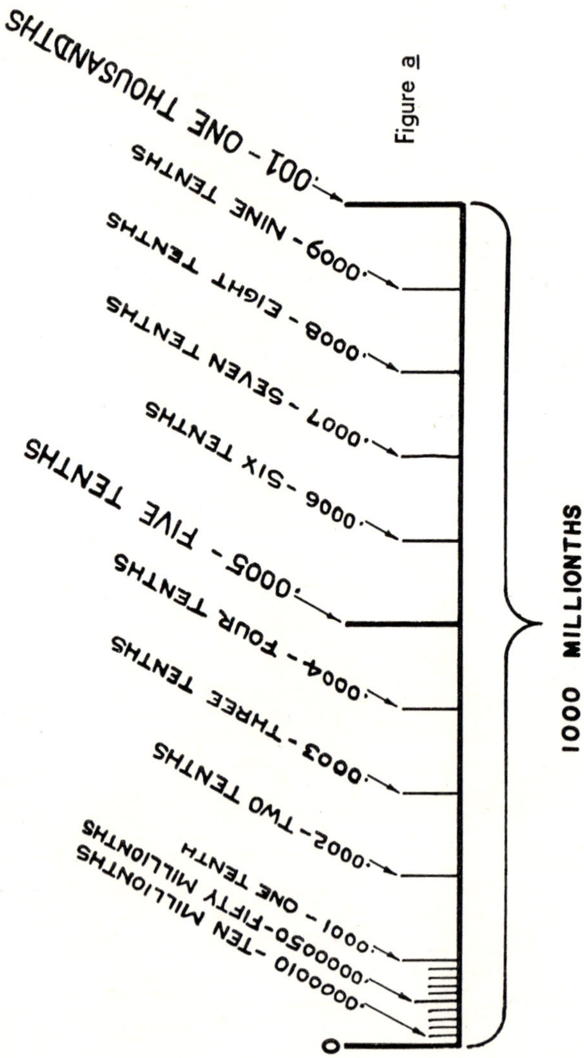

Figure a

NOMENCLATURE OF THE CIRCLE

The parts of the circle (see Figure *a*) are as follows:

1. CIRCUMFERENCE—A curved line that bounds a circle.
2. DEGREE—A division or interval marked on a measuring or other instrument, as on a protractor. The 360th part of the circumference of a circle.
3. DIAMETER—A straight line drawn through the center and terminating at both ends in the circumference.
4. RADIUS—A straight line between the circumference and the center. The radii of the same circle are of equal length and one half of the diameter.
5. ARC—Any part of the circumference of a circle.
6. CHORD—A straight line other than a diameter drawn inside a circle between two points of the circumference.
7. SEGMENT—The area included between an arc and a chord.
8. SECTOR—The area included between an arc and two radii drawn to the ends of the arc.

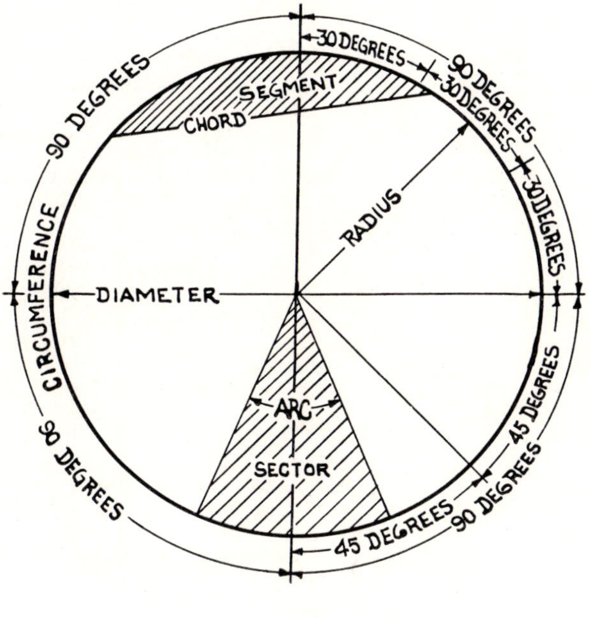

GEOMETRICAL FORMS AND THEIR MEANING

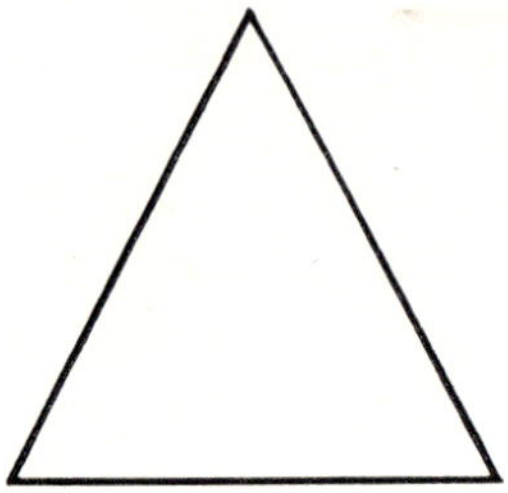

TRIANGLE

Figure a

TRIANGLE—a plane figure bounded by three lines and having three angles.

SQUARE

Figure b

SQUARE—a plane figure of four sides. All four angles are right angles; all sides are equal; and the opposite sides are parallel.

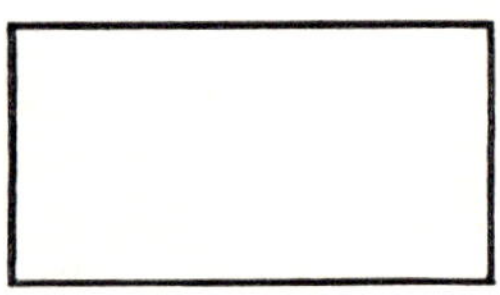

RECTANGLE

Figure c

RECTANGLE—a plane figure of four sides. All four angles are right angles. The adjacent sides need not be equal.

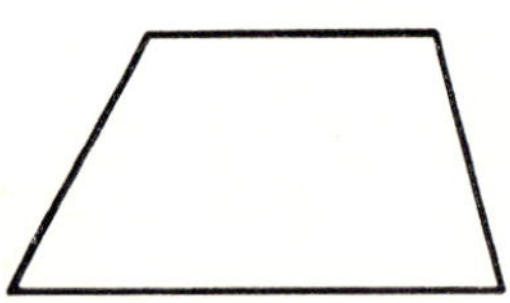

TRAPEZOID

Figure d

TRAPEZOID—a plane figure of four sides, two of which are parallel.

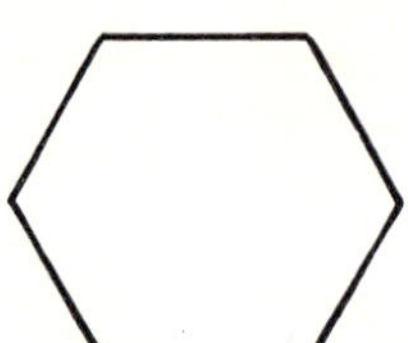

HEXAGON

Figure e

HEXAGON—a plane figure of six sides. All the sides are equal and all the angles are equal.

Ellipse—a plane figure bounded by a curve every point of which is the same distance from two points on the longest axis.

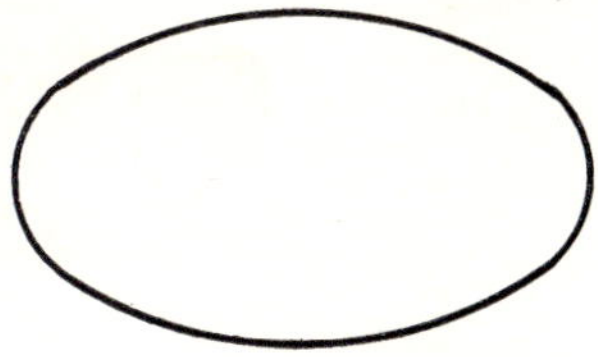

ELLIPSE

Figure f

Cube—a body having six faces, each a square. All faces and edges are equal.

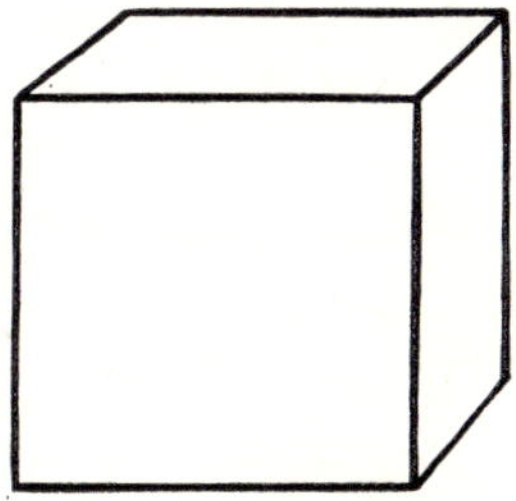

CUBE

Figure g

Pyramid—a body having a regular polygon as a base and isosceles triangles as faces.

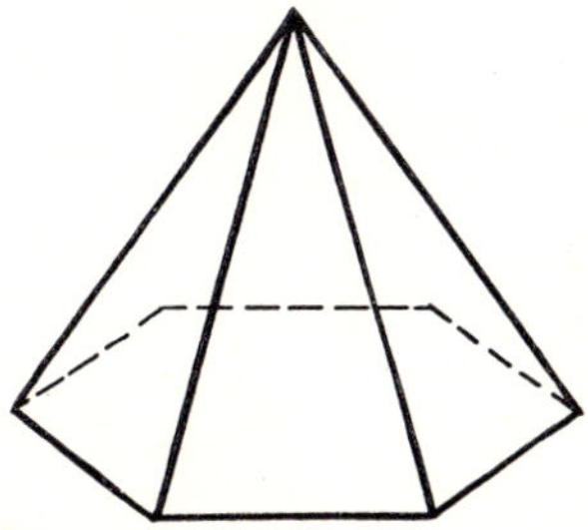

PYRAMID

Figure h

Cone—a body having a circular base and a vertex in a line perpendicular to the center of the base. It is formed by revolving a right triangle on its vertical side as an axis.

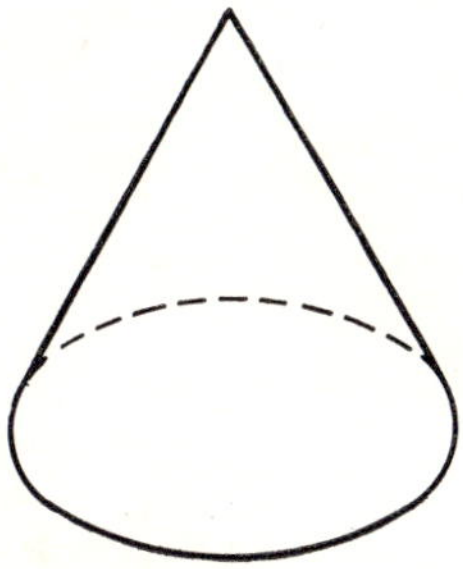

CONE

Figure i

CALCULATIONS

Right Triangles

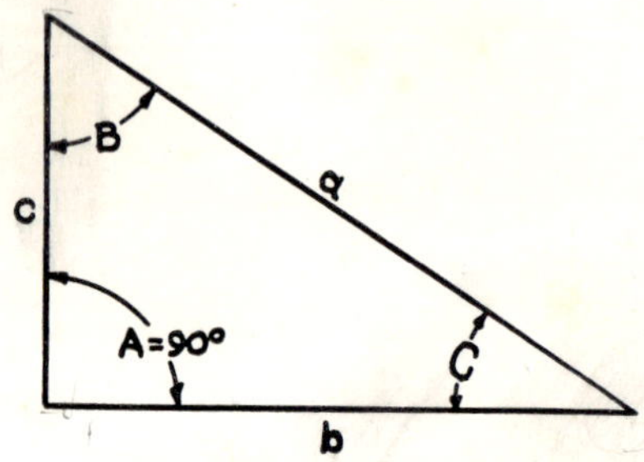

TO FIND ANGLES	FORMULAS	
C	$\frac{c}{a}$ = Sine C	90° − B
C	$\frac{b}{a}$ = Cosine C	90° − B
C	$\frac{c}{b}$ = Tangent C	90° − B
C	$\frac{b}{c}$ = Cotangent C	90° − B
C	$\frac{a}{b}$ = Secant C	90° − B
C	$\frac{a}{c}$ = Cosecant C	90° − B
B	$\frac{b}{a}$ = Sine B	90° − C
B	$\frac{c}{a}$ = Cosine B	90° − C
B	$\frac{b}{c}$ = Tangent B	90° − C
B	$\frac{c}{b}$ = Cotangent B	90° − C
B	$\frac{a}{c}$ = Secant B	90° − C
B	$\frac{a}{b}$ = Cosecant B	90° − C

TO FIND SIDES	FORMULAS	
a	$\sqrt{b^2 + c^2}$	
a	c × Cosecant C	$\frac{c}{\text{Sine C}}$
a	c × Secant B	$\frac{c}{\text{Cosine B}}$
a	b × Cosecant B	$\frac{b}{\text{Sine B}}$
a	b × Secant C	$\frac{b}{\text{Cosine C}}$
b	$\sqrt{a^2 - c^2}$	
b	a × Sine B	$\frac{a}{\text{Cosecant B}}$
b	a × Cosine C	$\frac{a}{\text{Secant C}}$
b	c × Tangent B	$\frac{c}{\text{Cotangent B}}$
b	c × Cotangent C	$\frac{c}{\text{Tangent C}}$
c	$\sqrt{a^2 - b^2}$	
c	a × Cosine B	$\frac{a}{\text{Secant B}}$
c	a × Sine C	$\frac{a}{\text{Cosecant C}}$
c	b × Cotangent B	$\frac{b}{\text{Tangent B}}$
c	b × Tangent C	$\frac{b}{\text{Cotangent C}}$

CALCULATIONS

Oblique Triangles

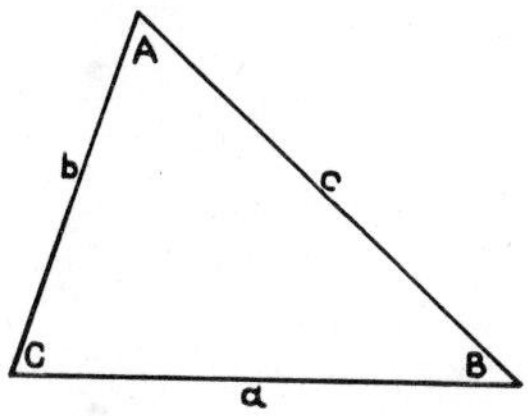

TO FIND	KNOWN	SOLUTION
C	A-B	$180° - (A + B)$
b	a-B-A	$\frac{a \times \text{Sine } B}{\text{Sine } A}$
c	a-A-C	$\frac{a \times \text{Sine } C}{\text{Sine } A}$
Tangent A	a-C-b	$\frac{a \times \text{Sine } C}{b - (a \times \text{Cosine } C)}$
B	A-C	$180° - (A + C)$
Sine B	b-A-a	$\frac{b \times \text{Sine } A}{a}$
A	B-C	$180° - (B + C)$
Cosine A	a-b-c	$\frac{b^2 + c^2 + a^2}{2bc}$
Sine C	c-A-a	$\frac{c \times \text{Sine } A}{a}$
Cotangent B	a-C-b	$\frac{a \times \csc C}{b}$ Cotangent C
c	b-C-B	$b \times \text{Sine } C \times \csc B$

Formulas for Finding Functions of Angles

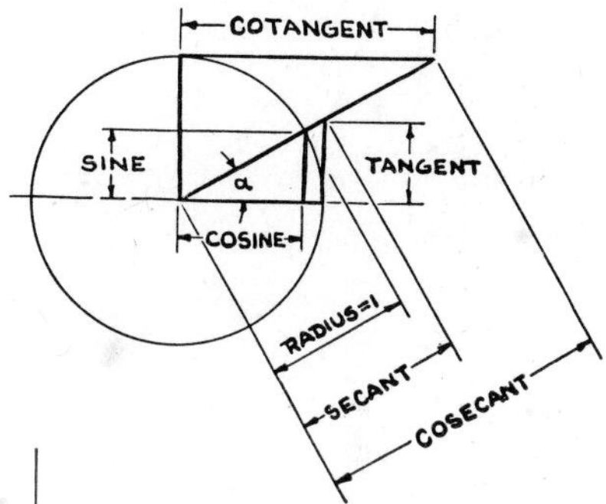

$\frac{\text{Side opposite}}{\text{Hypotenuse}}$ = SINE

$\frac{\text{Side adjacent}}{\text{Hypotenuse}}$ = COSINE

$\frac{\text{Side opposite}}{\text{Side adjacent}}$ = TANGENT

$\frac{\text{Side adjacent}}{\text{Side opposite}}$ = COTANGENT

$\frac{\text{Hypotenuse}}{\text{Side adjacent}}$ = SECANT

$\frac{\text{Hypotenuse}}{\text{Side opposite}}$ = COSECANT

FORMULAS FOR FINDING THE LENGTH OF SIDES FOR RIGHT-ANGLE TRIANGLES WHEN AN ANGLE AND SIDE ARE KNOWN

Length of side opposite:
- Hypotenuse × Sine
- Hypotenuse ÷ Cosecant
- Side adjacent × Tangent
- Side adjacent ÷ Cotangent

Length of side adjacent:
- Hypotenuse × Cosine
- Hypotenuse ÷ Secant
- Side opposite × Cotangent
- Side opposite ÷ Tangent

Length of hypotenuse:
- Side opposite × Cosecant
- Side opposite ÷ Sine
- Side adjacent × Secant
- Side adjacent ÷ Cosine

HOW TO CHECK INTERNAL AND EXTERNAL MILITARY PIPE THREADS

Military Specification Mil-P-7105; (formerly AN-P-363)

When taper pipe threads are made accurately and gaged properly, they will seal tightly without spiral leakage, without the use of a sealing compound, and without selective assembly.

Safety and interchangeability are important considerations when taper pipe threads are used in aviation because:

1. joints that can be shaken loose during flight may allow gasoline or oil to spray out and burn;
2. foreign substances, such as a sealing compound, may cause an explosion in an oxygen line;
3. compound creeping into an oil or gasoline line can stop or impede the flow of fuel or lubricant;
4. selective assembly is not only costly in time and labor, but a decided stumbling block in time of war, when interchangeable spare parts for aircraft are needed in every theater of combat.

For these reasons, the Army, the Air Force, and the Navy set up specifications for gaging pipe threads to meet certain safety requirements. These specifications are designated as Pipe Threads, Taper, Aeronautical National Form; symbol ANPT.

For checking the accuracy of internal threads, you need three gage members—a double-end thread plug, and a single-end plain taper plug. These are illustrated in Figure *a*. The L-1 member checks the diameter, lead, form, and taper of that portion of the thread which will be engaged normally when the part is screwed in by hand. The other end of this gage—the L-3 member—checks diameter, lead, form, and taper of threads farther in, that is, those threads which will be engaged by wrench pressure. The plain taper plug C checks the taper, the roundness, and also gives a close check on the diameter at the crest of the threads.

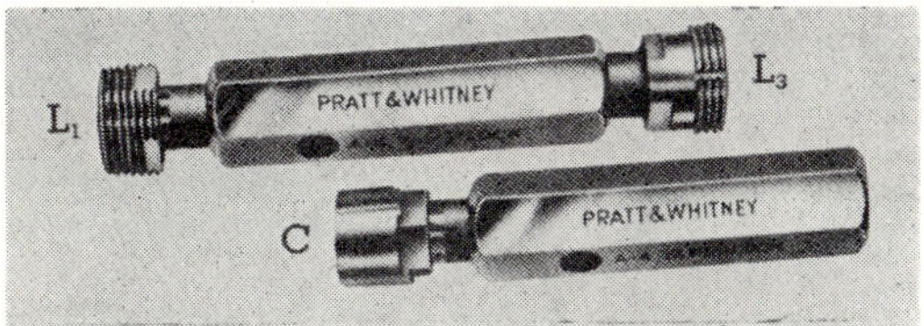

Figure a

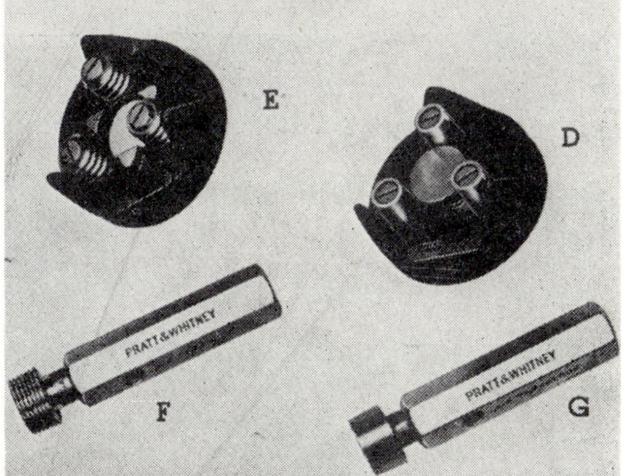

Figure b

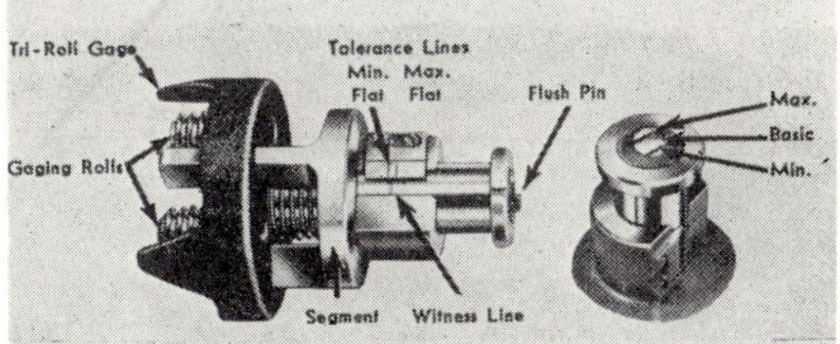

Figure c

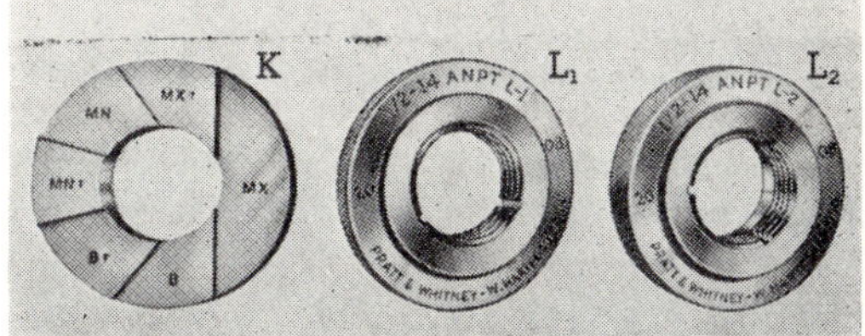

Figure d

EXTERNAL THREADS

To check external threads, you need two Tri-Roll gages with a setting plug for each.

As furnished, these two Tri-Roll gages D and E are correctly set (see Figure *b*). The two setting plugs, F and G are for checking the gages occasionally and resetting if necessary because of wear or damage. The Tri-Roll gage E checks the lead, taper, pitch diameter, and form of the thread. Three steps on the upper surface of the base give the operator the story. The Tri-Roll gage D checks the taper and the diameter of the threads at the crest. The part sits on a plunger which extends through the base where three steps are located for giving the operator a "reading."

Combination External

The combination Tri-Roll gage (see Figure *c*) combines the functions of gages D and E in Figure *b* into one operation. It checks simultaneously the thread size or pitch diameter and its relation to the permissible width of flat on the crest of the thread or the major diameter.

Alternate External

Specification MIL-P-7105 contains an alternate set of gages for checking external taper threads. These gages produce equally satisfactory results, but require more time than the Tri-Roll gages illustrated in Figure *c*. Either method is approved and may be adopted at the discretion of the user. The thick ring L-2 and the thin ring L-1 (see Figure *d*) are used to check the lead, diameter, taper, and form of the thread by first applying one, then the other. Then the plain taper ring K with six steps on its bottom is used to check the taper and diameter of threads at their crest (see Figure *d*).

STEP-BY-STEP CHECKING OF THREADS

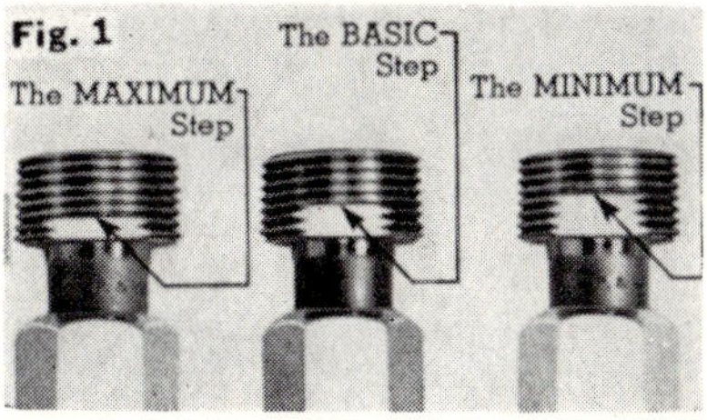

Pick up the double-end thread plug gage and find the end with the largest number of threads. This will be the L-1 gage shown in Figure *a* on page 81. Hold it in the position shown in Figure 1. Revolve it until you are looking at the narrowest flat place. This is the *maximum* step. Then revolve the gage until you have mentally identified all three steps as shown. To know these steps at a glance is important.

You are now ready to use the gage. But first *clean* the gage and *clean* the work. Remember that the best gage in the world will soon lose its accuracy if grit and dust are allowed to act as abrasives when the gage is moved in the work. Now screw gage L-1 into the work as far as it will go without forcing. When it stops, look to see where the steps come in relation to the edge of the hole in the work. The *minimum* step on the gage must at least reach the edge of the hole. If it does not—the thread is too small. The *maximum* step must not pass the edge of the hole as you screw the gage in. If it does—the thread is too large. If you find the threaded hole either too large or too small, the thread is not acceptable. But let us assume that the *minimum* step does reach the edge of the hole when the gage stops—as shown in Figure 2. This gives you the family name of this particular thread. From now on it is a *minimum* thread—and you work accordingly.

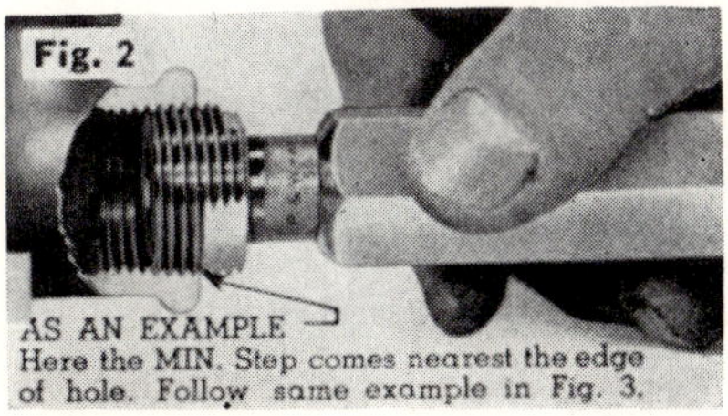

You are now ready to check the L-3 threads. Screw the L-3 plug (see Figure *a,* page 81) into the work as far as it will go without forcing. Now look at the steps. This plug has the same three steps as the first gage, or L-1. *Remember you have already found this to be a minimum thread.* (Figure 2.) If the thread is correct—the *minimum* step on the L-3 gage also will be nearest the hole, as illustrated in Figure 3. The relative position of the *minimum* steps on the two gages must not vary more than one-half turn. If more than one-half turn difference exists—the work will not pass. If the *minimum* step on the L-3 does come nearest the hole, and within one-half turn of the L-1, your thread is passable, up to this point, and you are ready to check the minor diameter, taper, and roundness.

Fig. 3

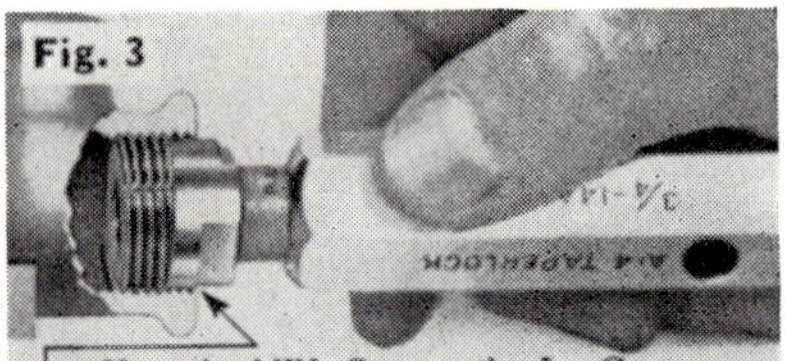

Here the MIN. Step on the L_3 Gage also comes nearest the edge of hole.

Note that the radial location of the relative gaging notches is not important. Only the variation from the face of the product is considered.

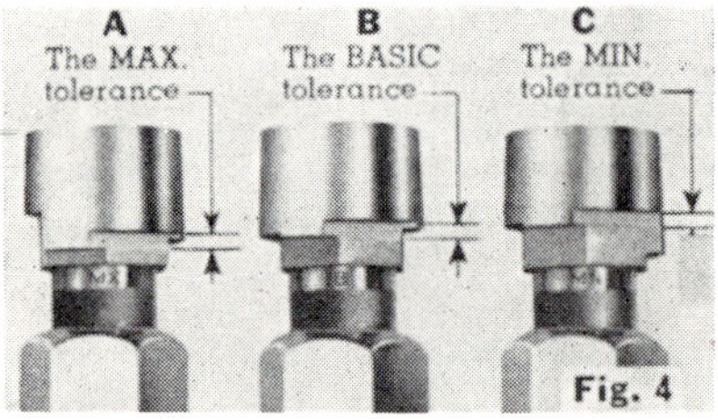

Fig. 4

Now you are ready to check the work for correct taper and minor diameter, that is, the inside diameter of the crests of the threads. Pick up the plain taper plug C (shown in Figure *a,* page 81) and examine it. It has six steps, instead of only three, as on the L-1 and L-3 gages. These six are divided into three pairs as shown in Figure 4. Hold the gage in your hand in the position shown and revolve it so that it looks like A in Figure 4. This is the tolerance for the *maximum.* Now revolve the gage to position B and locate the flat which gives you the tolerance for the *basic* step; then to position C which shows you the tolerance for the *minimum* step. Be sure you understand these flats before you attempt to use the gage.

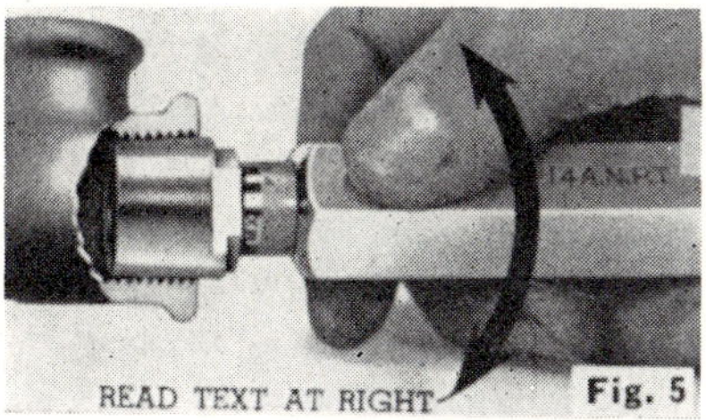

READ TEXT AT RIGHT Fig. 5

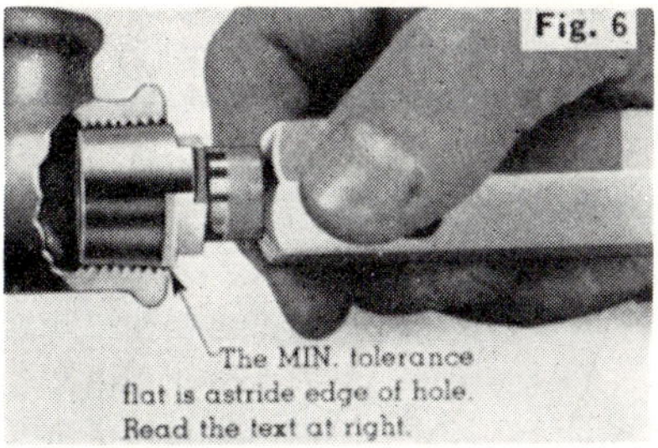

Fig. 6

The MIN. tolerance flat is astride edge of hole. Read the text at right.

Now place the taper plug in the hole as far as it will go without forcing. Try to shake it. Hold it to its full seat against the threads and try to move the handle up and down. If you feel excessive shaking, the work is not acceptable. (The taper tolerance on the product is plus and minus 1/16 inch per foot.)

Remember that you established this as a *minimum* thread with the gage L-1 in Figure 2. Your L-3 gage in Figure 3 also showed it to be a *minimum* thread. Now your taper plug must also show it to be a *minimum* if it is acceptable. So check your taper plug in the hole until you see which of the three flats comes even with, or straddles, the edge of the hole. If you check at the *minimum* flat as shown in Figure 6, then your thread and taper are acceptable. Once you start back at the L-1 stage with a *minimum* thread, the gaging operation must show up *minimum* all the way through. If, however, back in the L-1 gaging operation, your thread shows up *maximum,* then your L-3 and the plain taper plug must also show the hole to be *maximum.* Otherwise the work is rejectable. Of course, the same applies to the *basic* step.

For checking external ANPT threads a Tri-Roll thread gage (see Figure *b* on page 81) is used, as shown in Figure 7. This has three steps clearly marked *minimum, basic* and *maximum.* Pick up your gage and study it. Note one feature which is very different from the L-1 and L-3 gages which you used in Figures 2 and 3. There, the *minimum* was the highest step. On the Tri-Roll thread gage the *minimum* is the lowest step. Watch this difference.

Screw the part into the tri-rolls as far as it will go without forcing. The bottom of the part must screw down to the highest step. If it does not, it is too large. If it screws down past the lowest step, it is too small. In either case, it is rejected. Now look to see which step is nearest the bottom of the part. In Figure 7, the *basic* step happens to come just flush with the bottom of the part. This establishes it as a *basic* thread. You are now ready to check this part for taper and diameter at the crest of the threads.

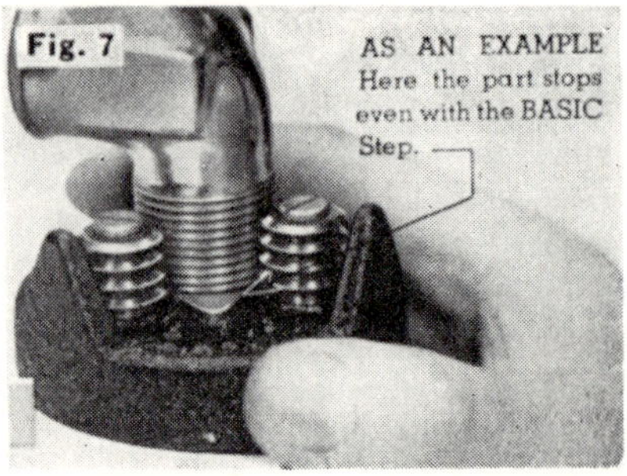

To check the taper and diameter, you use the plain taper Tri-Roll gage as shown in Figure *b* (page 81). This has three marked steps on the bottom and a stepped plunger which runs through and reaches these steps. The step on the plunger is the tolerance. Place the part on the spring plunger and push it down until it seats or stops. It is then touching all three taper rolls. Do not use force. At this point, you make a visual check of the taper by sighting between the rolls and the thread. Now hold the part down against the plunger with the fingers (see Figure 8).

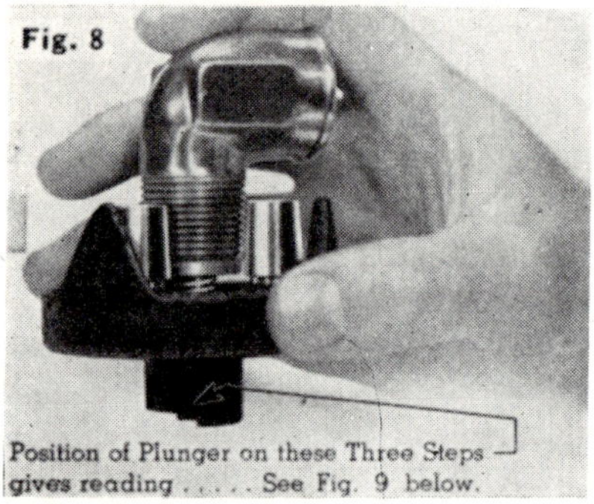

When you first glance at the steps it may look confusing, but it will be easy to recognize the correct reading if you remember that you have to take into consideration only one of the three steps on the bottom of the gage. *In this case your thread has already been identified as basic* (Figure 7). You look to see if the step on the plunger straddles or comes even with the edge of the *basic* step on the bottom of the gage. If it does, as shown in Figure 9, your part is acceptable. If it does not, the part is not acceptable.

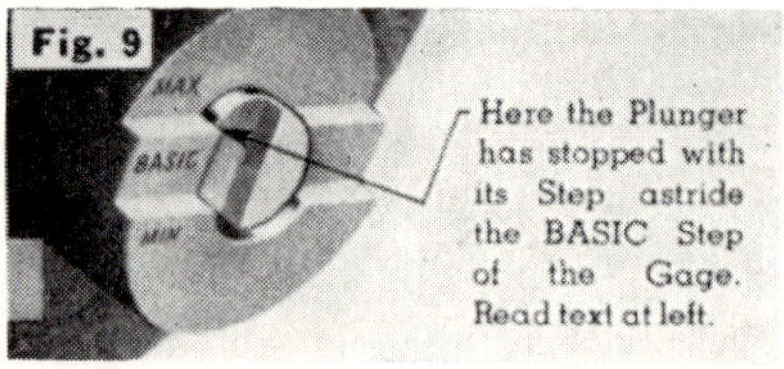

The same holds true, of course, if your Tri-Roll thread gage (Figure 7) established this thread to be a *maximum*. Then you will be interested only in the *maximum* step on the bottom of the plain taper gage. The same considerations apply to the *minimum*.

THE ALTERNATE METHOD FOR CHECKING EXTERNAL THREADS

Fig. 10

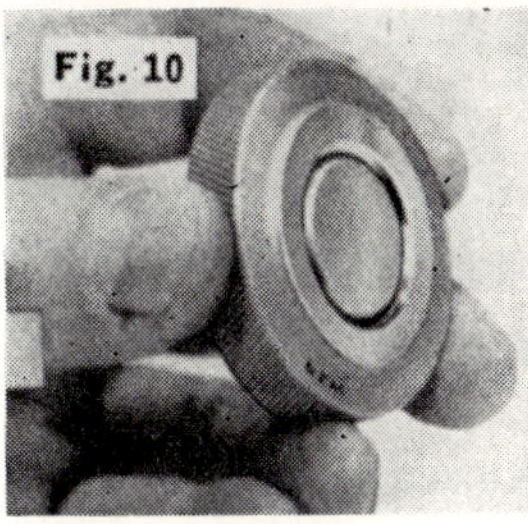

As an EXAMPLE Here the part stops flush with end of Ring Gage, hence is BASIC.

For checking external threads by the alternate method, you need two thread ring gages and a six-step plain taper ring gage (as shown in Figure *d,* on page 81). There are no steps or marks on the two thread ring gages to show *minimum, basic,* and *maximum.* You must determine these by watching the position of the part in the gage.

Screw the thick L-2 ring (see Figure *d* on page 81) on the work as far as it will go without using force. If it stops flush with the end of the part, as in Figure 10, it is a *basic* thread. Let us assume that your thick ring gage shows your part to be *basic,* as in Figure 10.

Fig. 11

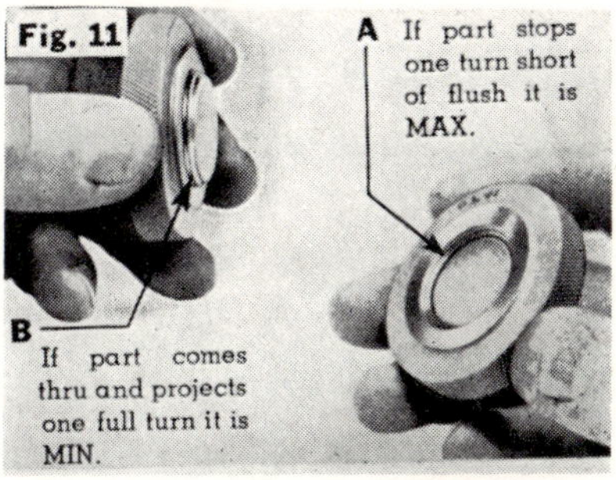

If the part stops one turn short of being flush with the end of the gage, as shown in A, it is *maximum.* If it stops more than one turn short of coming up to flush, it is too large and will not pass. If it comes all the way through and projects one full turn as at B, it is *minimum.* If it projects through the gage more than one turn, it is too small and not acceptable. After you have decided the family (or classification) of your part whether it is *maximum, basic,* or *minimum,* you are ready to use the second or thin ring gage. Remove the thick ring L-2 and screw on the thin ring L-1 (see Figure *d* on page 81) as far as it will go. Determine what classification it gives to the thread. It must give you a "reading" to match within one-half turn the results which the first or thick ring gage showed. If your thick ring gage showed the thread to be *basic,* as in Figure 10, then the thin ring gage must also show it to be *basic.* If either thread ring gage goes on to the part more than one-half turn farther than the other, the piece is not acceptable.

Fig. 12

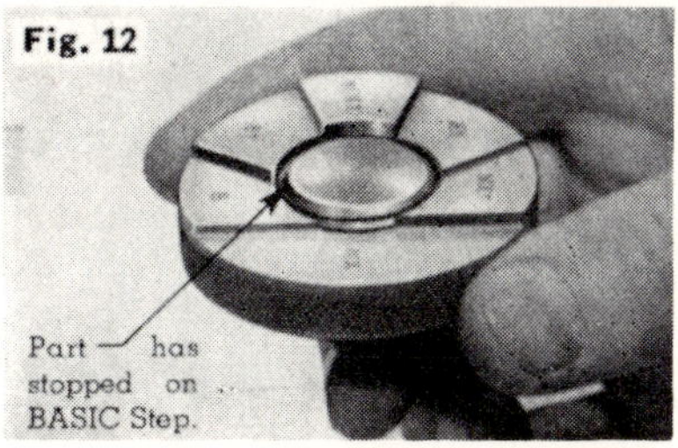

Now place the part into the taper ring K of Figure *d* (on page 81) and try it for taper. "Off taper" will show up quickly by side shake or wobble of the part in the hole. Now look at the steps on the bottom. If it is a *basic* thread (as established in Figure 10), then the bottom of the part must stop somewhere between the *basic* and *basic tolerance* steps, as shown in Figure 12. If it does not, throw it out. The same goes for a *maximum* thread. It must stop between the *maximum* and *maximum tolerance* steps. The same condition is true of a *minimum* thread.

DIAMETRAL PITCH or D.P. CALCULATIONS

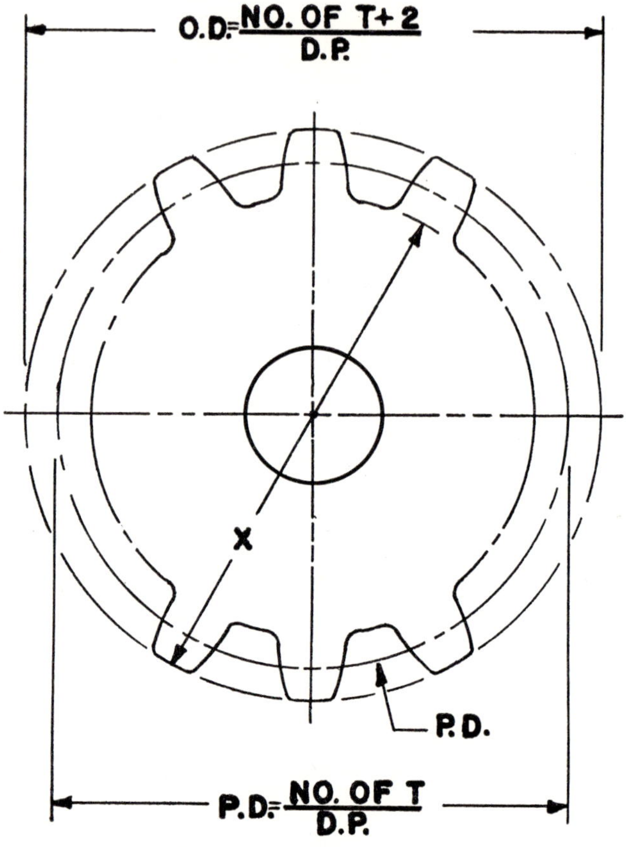

Figure *a*

Diametral pitch or tooth size number is the number of teeth for each inch of a pitch diameter. For example: 12 DP, 1 inch pitch diameter has twelve teeth. The pitch diameter (PD) of a gear is considered to be the diameter of a gear, and not the outside diameter (OD). The diametral pitch can be found by counting the number of teeth and dividing them by the pitch diameter. The approximate pitch diameter can be measured from the bottom of one tooth to the top of the one diagonally opposite it, see X shown in Figure *a*.

Example: 20 Teeth, 4 inch Diameter $= \frac{20}{4} = 5$ DP.

Complete diametral pitch formulas are shown on the facing page.

DIAMETRAL PITCH FORMULAS

Rule No.	Dimension Wanted	Rule	Formula
1	Circular Pitch (CP)	Divide 3.1416 by Diametral Pitch........ ..	$CP = \frac{3.1416}{DP}$
2	Circular Pitch (CP)	Multiply Pitch Diameter by 3.1416 and divide Product by Number of Teeth.............	$CP = \frac{PD \times 3.1416}{N}$
3	Diametral Pitch (DP)	Divide 3.1416 by Circular Pitch........	$DP = \frac{3.1416}{CP}$
4	Diametral Pitch (DP)	Divide Number of Teeth by Pitch Diameter.....	$DP = \frac{N}{PD}$
5	Pitch Diameter (PD)	Divide Number of Teeth by Diametral Pitch ..	$PD = \frac{N}{DP}$
6	Pitch Diameter (PD)	Multiply Number of Teeth by Circular Pitch and divide Product by 3.1416.............	$PD = \frac{N \times CP}{3.1416}$
7	Center Distance (CDi)	Add Number of Teeth in Gear and Pinion, and divide Sum by Twice Diametral Pitch........	$CDi = \frac{Ng + Np}{2DP}$
8	Center Distance (CDi)	Multiply Sum of Number of Teeth in Gear and Pinion by Circular Pitch, and divide Product by 6.2832...........................	$CDi = \frac{(Ng + Np)CP}{6.2832}$
9	Addendum (A).......	Divide 1 by Diametral Pitch............... (For Full-length Teeth)	$A = \frac{1}{DP}$
10	Addendum (A).......	Divide Circular Pitch by 3.1416............ (For Full-length Teeth)	$A = \frac{CP}{3.1416}$
11	Clearance (C)		See Tables IV and V
12	Dedendum(D)	Add Clearance to Addendum...............	$D = A + C$
13	Whole Depth (WD).....	Add Dedendum to Addendum...............	$WD = A + D$
14	Circular Thickness (CTh).	Divide 1.5708 by Diametral Pitch...........	$CTh = \frac{1.5708}{DP}$
15	Circular Thickness (CTh)	Divide Circular Pitch by 2.................	$CTh = \frac{CP}{2}$
16	Outside Diameter (OD)	Add 2 to Number of Teeth, and divide Sum by Diametral Pitch......................... (For Full-length Teeth)	$OD = \frac{N + 2}{DP}$
17	Outside Diameter (OD)	Add Two Times the Addendum to the Pitch Diameter............................	$OD = PD + 2A$
18	Number of Teeth (N)	Multiply Pitch Diameter by Diametral Pitch...	$N = PD \times DP$
19	Number of Teeth (N)	Multiply Pitch Diameter by 3.1416, and divide Product by Circular Pitch................	$N = \frac{PD \times 3.1416}{CP}$
20	Root Diameter (RD)	Subtract Two Times the Whole Depth from Outside Diameter.........................	$RD = OD - 2WD$
21	Base Circle Diameter ($BCDi$)....	Multiply Pitch Diameter by Cosine of Pressure Angle..............................	$BCDi = PD \times CosVP$

INTERPRETATION OF MECHANICAL DRAWINGS

A drawing or blue print is a device by which any information concerning the size and shape of an item is relayed to the mechanic in a visual manner.

To interpret a drawing accurately, one must understand lines and symbols which the draftsman uses to represent various details and one must be able to visualize the complete object on examining the drawing.

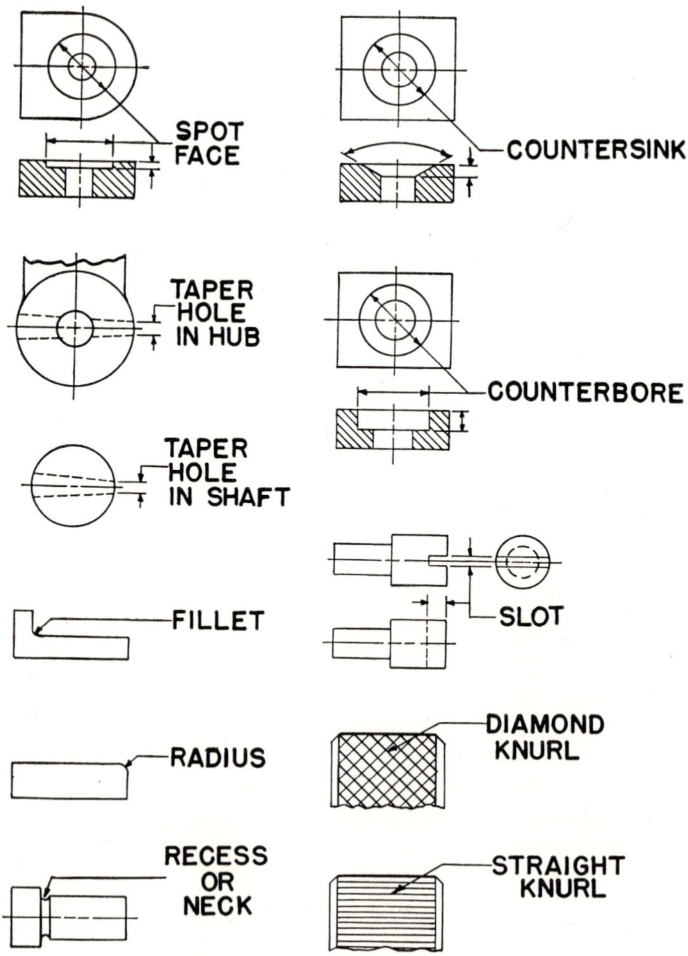

MECHANICAL DRAWING SYMBOLS

In mechanical drafting, a system of symbols is used that helps eliminate written instructions for the manufacture of an article or machine.

Figure *a* illustrates the common symbols used in standard drafting practice.

Whenever a part appears on a drawing with a break through it and cross sectioned as shown in Figure *b*, it does not mean that the part is to be made like this. The reason for the break is that the part was drawn to scale and since its length was greater than the space available on the drawing paper, the draftsman drew a break through the part as shown.

The same method is used for illustrating long parts of square, hexagonal, or any other shape.

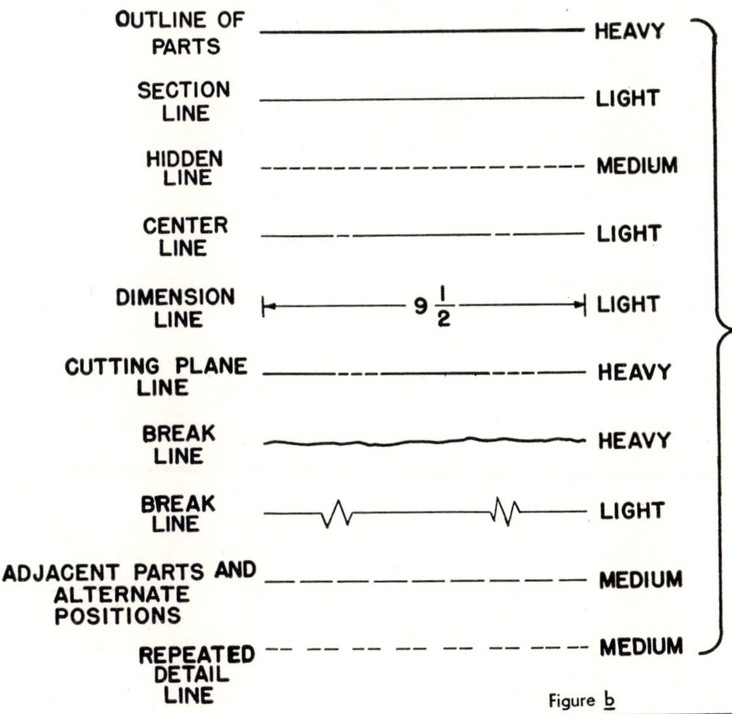

Figure a

Figure b

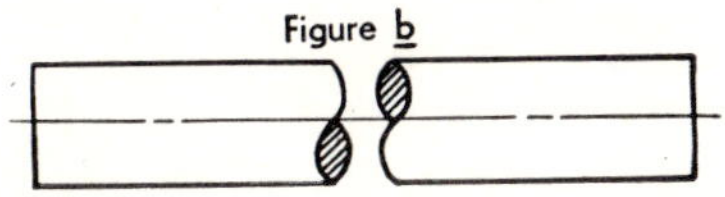

INDICATION OF MATERIALS

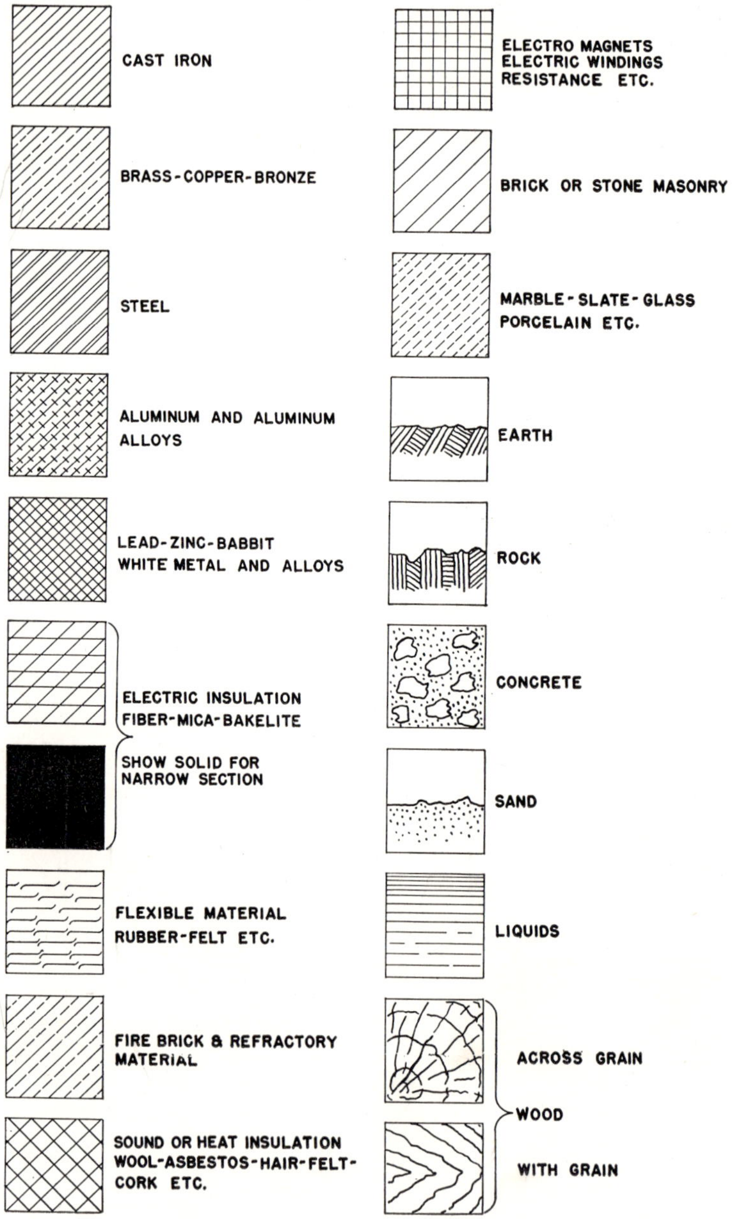

INDICATION OF DIRECTION

Assembly and subassembly drawings often show with arrows in which direction a shaft, lever, ram, carriage, etc., must move when properly assembled.

Each of Figures *a* to *i* shows an arrow which indicates the direction of motion required and explains the meaning of the arrow.

Figure a

ROTATING TOWARD OPERATOR

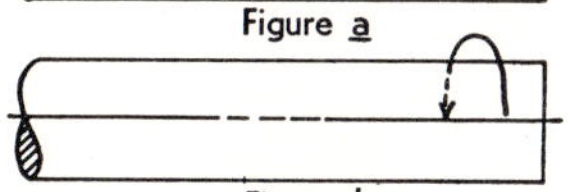

Figure b

ROTATING AWAY FROM OPERATOR

Figure c

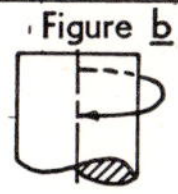

ROTATING TO THE RIGHT

Figure d

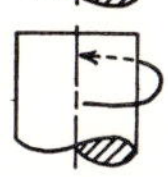

ROTATING TO THE LEFT

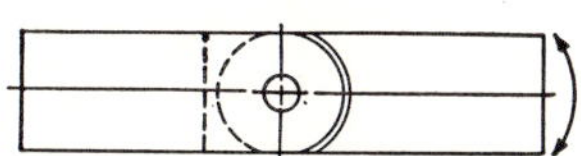

Figure e

CIRCULAR MOVEMENT UP AND DOWN

Figure f

CIRCULAR MOVEMENT DOWNWARD

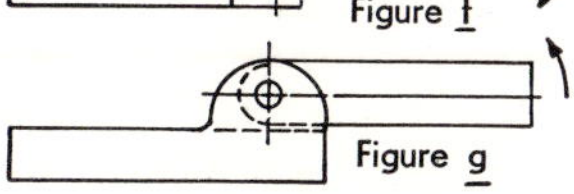

Figure g

CIRCULAR MOVEMENT UPWARD

Figure h

VERTICAL MOVEMENT

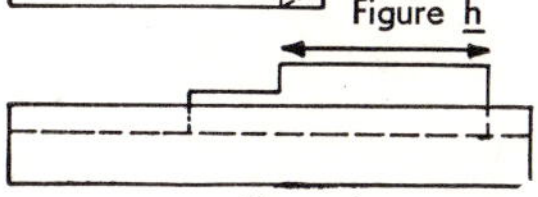

Figure i

HORIZONTAL MOVEMENT

MISCELLANEOUS SYMBOLS

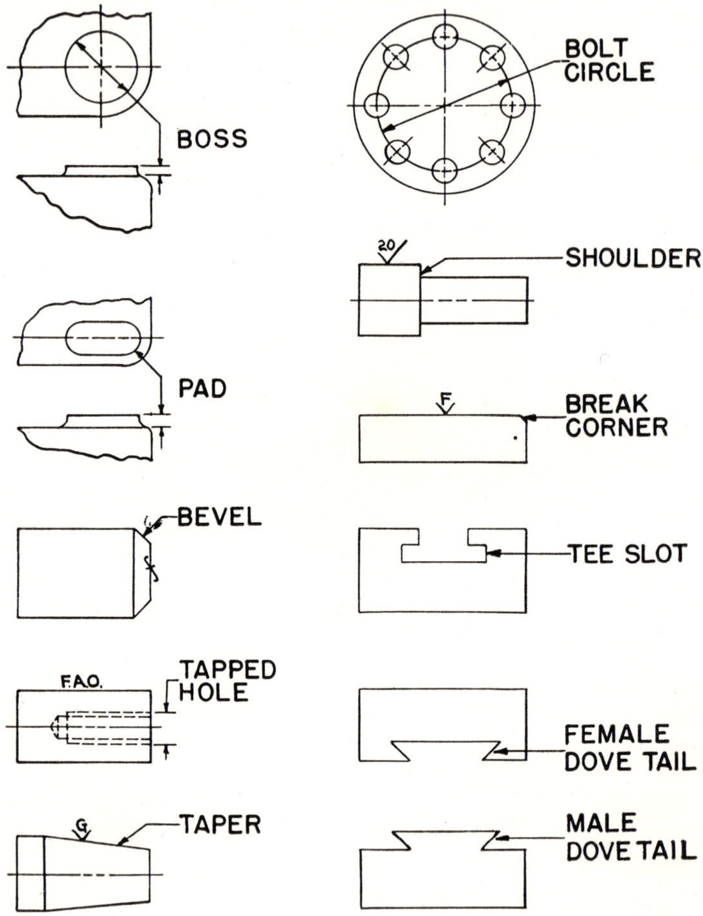

$\overset{20}{\surd}$—Indicates *micro inch* finish.
F—Indicates *file* finish.
G—Indicates *ground* finish.
f—Indicates *machine* finish.
F.A.O.—Indicates *finish all over.*

ABBREVIATIONS FREQUENTLY USED ON DRAWINGS

A.S.A.—American Standards Association

B.B.—Ball Bearings

B.C.—Bolt Circle

Bab.—Babbit Metal

Br.—Brass

Bz. or Bro.—Bronze

C'bore—Counterbore

C.I.—Cast Iron

Cop.—Copper

C.P.—Circular Pitch

C.R.S.—Cold-Rolled Steel

Csk.—Countersink

CtoC—Center to Center

D—Diameter

D.F.—Drop Forging

D.P.—Diametral Pitch

EF—Extra Fine

f.—Finish

Fil.—Filister

G.I.—Galvanized Iron

Hd.—Head

Hex.—Hexagonal

Hrd. & Gr.—Harden and Grind

L.H.—Left Hand

Mal. I.—Malleable Iron

M.S.—Machine Steel

No. or #—Number

O.D.—Outside Diameter

P.—Pitch

Pat.—Pattern

Pcs.—Pieces

P.D.—Pitch Diameter

Phos. Bro.—Phosphor Bronze

lb. or #—Pound

R.—Radius

Rd.—Round

Req.—Required

R.H.—Right Hand

S. or St.—Mild Steel

S.C.—Steel Casting

Sc.—Screw

S. Forg.—Steel Forging

Std.—Standard

S. Tube—Steel Tubing

T.—Teeth

Thd.—Thread

T.S.—Tool Steel

W.I.—Wrought Iron

STANDARD SYSTEM OF MARKING AMERICAN THREAD

Symbol	Reference
NC	American National Coarse Thread Series
NF	American National Fine Thread Series
NEF	American National Extra Fine Thread Series
N	American National 8, 12, and 16 Pitch Series (8N, 12N, 16N)
NS	Special Threads of American National
NH	American National Hose Coupling and Fire-Hose Coupling Thread
NPT	American Standard Taper Pipe Thread
NPTF	American Standard Taper Pipe Thread (Dryseal)
NPTR	American Standard Taper Pipe Thread for Railing Fittings
ANPT	Military Pipe Thread Specification MIL–P–7105
NPS	American Standard Straight Pipe Thread
NPSC	American Standard Straight Pipe Thread for Pipe Couplings
NPSF	American Standard International Straight Pipe Thread (Dryseal)
NPSH	American Standard Straight Pipe Thread for Hose Coupling and Nipples
NPSI	American Standard Intermediate Internal Straight Pipe Thread (Dryseal)
NPSL	American Standard Straight Pipe Thread for Locknuts and Locknut Pipe Thread
NPSM	American Standard Straight Pipe Thread for Mechanical Joints
ACME	Acme Threads; Acme-C Centralizing; Acme-G General Purpose
STUB-ACME	Stub Acme Threads
NGO	American National Gas Outlet Thread
V	A60 V thread with truncated crest and roots; the theoretical V form is usually flatted several thousandths of an inch to the user's specifications.
SB	Manufacturers Stovebolt Standard Thread
	Symbols Used For Unified Threads:
UNC	Unified Coarse Thread Series
UNF	Unified Fine Thread Series
UNEF	Unified Extra Fine Thread Series

TRUNCATED AND FULL-FORM THREAD-SETTING PLUG GAGES

The truncated setting plug gage is recommended for setting and checking thread ring gages. These GO and NOT GO setting plug members have a truncated major diameter for approximately one-half length of the gage. The remaining half is full form (see Figure *a*).

The pitch diameter is constant for the entire thread length of the member. In using the truncated setting plug, the ring gage should be adjusted to fit the full portion; then the fit of the gage on the truncated portion should be determined. If the thread form of the ring gage is satisfactory, there will be a snug fit with very slight or no shake or play.

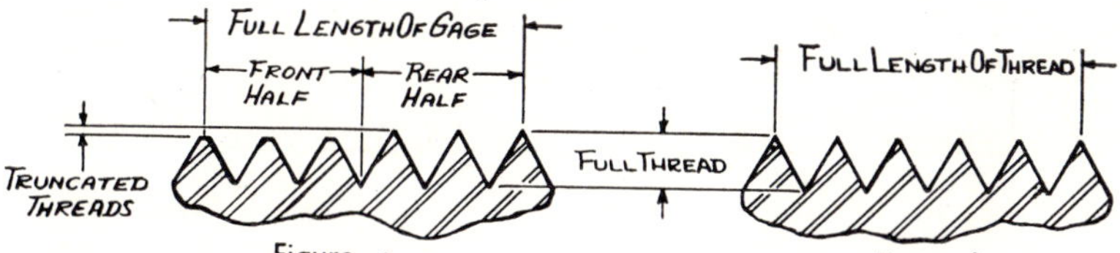

Figure a

Figure b

If the thread ring gage is worn, the presence of shake or play on the truncated portion indicates that the sides of the thread are no longer straight near the root and the gage should be relapped or discarded.

Full-form setting plugs are used for setting and checking roll thread snap gages and thread ring gages.

GO and NOT GO members have a full-form major diameter for the entire length (see Figure *b*).

GAGE CARE

Fixed Gages

1. Wipe gages to clean before using.
2. Use a gage only for its intended purpose.
3. Any surface that is to be gaged must be wiped clean.
4. Gages should never come into contact with tools of hard surface.
5. Keep gages away from excessive heat, hot sunshine, and chilling drafts.
6. Gages that have been dropped or struck should be checked before using. Reset if necessary.
7. Clean gages and place in proper containers after they are used to prevent them from rust or being damaged.

Air Gages

1. Examine orifices to make sure that they are not clogged.
2. Examine hoses for kinks.
3. Examine reducing valve and pressure gages.
4. Examine traps and oil filters for oil or water.

Electric Gages

1. Use only on current specified.
2. Check cycles, voltage, whether ac or dc.
3. Replace wiring if insulation becomes damaged.
4. Periodically check gages against standards.

APPENDIX

Constants for Finding Pitch Diameter and Minor Diameter of 60° Screw Threads

To find the pitch diameter or minor diameter of any 60° screw thread, subtract the constant for the number of threads per inch from the major diameter.

Threads per Inch	Pitch	Constants for Finding Pitch Diameter		Constants for Finding Minor Diameter		
		American National Thread	Theoretical V	U. S. Thread	Am. Nat'l. Thread Ring Gages	Theoretical V
80	.01250	.00812	.01083	.01624	.01353	.02166
72	.01389	.00902	.01203	.01804	.01504	.02406
64	.01563	.01015	.01353	.02030	.01692	.02706
60	.01667	.01083	.01443	.02165	.01804	.02887
56	.01786	.01160	.01546	.02320	.01933	.03093
50	.02000	.01299	.01732	.02598	.02165	.03464
48	.02083	.01353	.01804	.02706	.02255	.03608
44	.02273	.01476	.01968	.02952	.02460	.03936
40	.02500	.01624	.02165	.03248	.02706	.04330
36	.02778	.01804	.02406	.03608	.03007	.04811
32	.03125	.02030	.02706	.04059	.03383	.05413
30	.03333	.02165	.02887	.04330	.03609	.05773
28	.03571	.02320	.03093	.04639	.03866	.06186
27	.03704	.02406	.03208	.04812	.04010	.06416
26	.03846	.02498	.03331	.04996	.04164	.06662
24	.04167	.02706	.03608	.05413	.04511	.07217
22	.04545	.02952	.03936	.05905	.04921	.07873
20	.05000	.03248	.04330	.06495	.05413	.08660
18	.05556	.03608	.04811	.07217	.06014	.09623
16	.06250	.04059	.05413	.08119	.06766	.10825
14	.07143	.04639	.06186	.09279	.07732	.12372
13	.07692	.04996	.06662	.09993	.08327	.13323
12	.08333	.05413	.07217	.10825	.09021	.14434
11½	.08696	.05648	.07531	.11296	.09413	.15062
11	.09091	.05905	.07873	.11809	.09841	.15746
10	.10000	.06495	.08660	.12990	.10825	.17321
9	.11111	.07217	.09623	.14434	.12028	.19245
8	.12500	.08119	.10825	.16238	.13532	.21651
7	.14286	.09279	.12372	.18558	.15465	.24744
6	.16667	.10825	.14434	.21651	.18042	.28868
5½	.18182	.11809	.15746	.23619	.19683	.31492
5	.20000	.12990	.17321	.25981	.21651	.34641
4½	.22222	.14434	.19245	.28868	.24056	.38490
4	.25000	.16238	.21651	.32476	.27064	.43301
3½	.28571	.18558	.24744	.37115	.30930	.49487
3¼	.30769	.19985	.26647	.39970	.33309	.53294
3	.33333	.21651	.28868	.43301	.36085	.57733
2¾	.36364	.23619	.31492	.47238	.39368	.62984
2½	.40000	.25981	.34641	.51962	.43305	.69282
2¼	.44444	.28868	.38490	.57736	.48117	.76980
2	.50000	.32476	.43301	.64952	.54131	.86602

Wilson Chart 38: Approximate Relationships between Hardness Values Determined on Rockwell and Rockwell Superficial Hardness Testers and those of other Testers.

All so-called "conversion" tables of hardness scales, including those here published, are and must be based on the assumption that the metal tested is homogeneous to a depth several times as great as the depth of the indentation, because different loads and different shapes of penetrators would, in metal not homogeneous, penetrate—or at least meet the resistance of—metal of varying hardness, depending upon the depth of the indentation, hence, no definite hardness value would actually exist and no recorded hardness value would be valid to an extent that could be confirmed by another person unless shape of penetrator and actual load applied are both specified.

These fundamentally essential specifications as to both loads and penetrators are and always have been fully covered in the recording of readings obtained on "Rockwell" Hardness Testers by the use of alphabetical symbols.

Conversion tables dealing with hardness can be only approximate and never mathematically exact, for it must be understood that a penetration hardness test proceeds until the specimen tested supports the applied load. Yet it is a severely cold-worked metal that actually supports the penetrator, and different metals, different alloys, and different analyses of the same type of alloy, have different cold-working properties. Nevertheless, while a conversion table cannot be mathematically exact, it is of considerable value to be able to compare different hardness scales in a general way.

For unhardened steel, steel of soft temper, gray and malleable cast iron and most non-ferrous metal

B.	F.	G.	15-T **	30-T **	45-T **	E.	K.	A.	Bri-nell	Bri-nell	Tensile Strength	B.	F.	G.	15-T **	30-T **	45-T **	E.	H.	K.	A.	Bri-nell
"ROCKWELL" HARDNESS TESTER B scale 1/16″ Ball penetrator—100 kg Load	"ROCKWELL" HARDNESS TESTER F scale 1/16″ Ball penetrator—60 kg Load	"ROCKWELL" HARDNESS TESTER G scale 1/16″ Ball penetrator—150 kg Load	"ROCKWELL" SUPERFICIAL 15-T scale 1/16″ Ball penetrator—15 kg Load	"ROCKWELL" SUPERFICIAL 30-T scale 1/16″ Ball penetrator—30 kg Load	"ROCKWELL" SUPERFICIAL 45-T scale 1/16″ Ball penetrator—45 kg Load	"ROCKWELL" HARDNESS TESTER E scale 1/8″ Ball penetrator—100 kg Load	"ROCKWELL" HARDNESS TESTER K scale 1/8″ Ball penetrator—150 kg Load	"ROCKWELL" HARDNESS TESTER A scale "BRALE" penetrator—60 kg Load	Brinell, Standard Type, 500 kg Load 10 mm Ball penetrator	Brinell, Standard Type, 3000 kg Load 10 mm Ball penetrator	Thousand lbs per sq in.	"ROCKWELL" HARDNESS TESTER B scale 1/16″ Ball penetrator—100 kg Load	"ROCKWELL" HARDNESS TESTER F scale 1/16″ Ball penetrator—60 kg Load	"ROCKWELL" HARDNESS TESTER G scale 1/16″ Ball penetrator—150 kg Load	"ROCKWELL" SUPERFICIAL 15-T scale 1/16″ Ball penetrator—15 kg Load	"ROCKWELL" SUPERFICIAL 30-T scale 1/16″ Ball penetrator—30 kg Load	"ROCKWELL" SUPERFICIAL 45-T scale 1/16″ Ball penetrator—45 kg Load	"ROCKWELL" HARDNESS TESTER E scale 1/8″ Ball penetrator—100 kg Load	"ROCKWELL" HARDNESS TESTER H scale 1/8″ Ball penetrator—60 kg Load	"ROCKWELL" HARDNESS TESTER K scale 1/8″ Ball penetrator—150 kg Load	"ROCKWELL" HARDNESS TESTER A scale "BRALE" penetrator—60 kg Load	Brinell, Standard Type, 500 kg Load 10 mm Ball penetrator
100		82.5	93.0	82.0	72.0			61.5	201	240	116	50	85.5	2.5	77.0	49.5	23.0	87.0		64.5	35.0	83
99		81.0	92.5	81.5	71.0			61.0	195	234	112	49	85.0	1.0	76.5	49.0	22.0	86.5		63.5		82
98		79.0		81.0	70.0			60.0	189	228	109	48	84.5			48.5	20.5	85.5		62.5	34.5	81
97		77.5	92.0	80.5	69.0			59.5	184	222	106	47	84.0		76.0	47.5	19.5	85.0		61.5	34.0	80
96		76.0		80.0	68.0			59.0	179	216	103	46	83.0		75.5	47.0	18.5	84.5		61.0	33.5	
95		74.0	91.5	79.0	67.0			58.0	175	210	101	45	82.5			46.0	17.5	84.0		60.0	33.0	79
94		72.5		78.5	66.0			57.5	171	205	98	44	82.0		75.0	45.5	16.5	83.5		59.0	32.5	78
93		71.0	91.0	78.0	65.5			57.0	167	200	96	43	81.5		74.5	45.0	15.5	82.5		58.0	32.0	77
92		69.0	90.5	77.5	64.5		100	56.5	163	195	93	42	81.0			44.0	14.5	82.0		57.5	31.5	76
91		67.5		77.0	63.5		99.5	56.0	160	190	91	41	80.5		74.0	43.5	13.5	81.5		56.5	31.0	75
90		66.0	90.0	76.0	62.5		98.5	55.5	157	185	89	40	79.5		73.5	43.0	12.5	81.0		55.5		
89		64.0	89.5	75.5	61.5		98.0	55.0	154	180	87	39	79.0			42.0	11.0	80.0		54.5	30.5	74

88		62.5		75.0	60.5		97.0	54.0	151	176	85	38	78.5	73.0	41.5	10.0	79.5		54.0	30.0	73
87		61.0	89.0	74.5	59.5		96.5	53.5	148	172	83	37	78.0	72.5	40.5	9.0	79.0		53.0	29.5	72
86		59.0	88.5	74.0	58.5		95.5	53.0	145	169	81	36	77.5		40.0	8.0	78.5	100	52.0	29.0	
85		57.5		73.5	58.0		94.5	52.5	142	165	80	35	77.0	72.0	39.5	7.0	78.0	99.5	51.5	28.5	71
84		56.0	88.0	73.0	57.0		94.0	52.0	140	162	78	34	76.5	71.5	38.5	6.0	77.0	99.0	50.5	28.0	70
83		54.0	87.5	72.0	56.0		93.0	51.0	137	159	77	33	75.5		38.0	5.0	76.5		49.5		69
82		52.5		71.5	55.0		92.0	50.5	135	156	75	32	75.0	71.0	37.5	4.0	76.0	98.5	48.5	27.5	
81		51.0	87.0	71.0	54.0		91.0	50.0	133	153	74	31	74.5		36.5	3.0	75.5	98.0	48.0	27.0	68
80		49.0	86.5	70.0	53.0		90.5	49.5	130	150	72	30	74.0	70.5	36.0	2.0	75.0		47.0	26.5	67
79		47.5		69.5	52.0		89.5	49.0	128	147		29	73.5	70.0	35.5	1.0	74.0	97.5	46.0	26.0	
78		46.0	86.0	69.0	51.0		88.5	48.5	126	144	Even for steel, Tensile Strength relation to hardness is inexact, unless determined for specific material.	28	73.0		34.5		73.5	97.0	45.0	25.5	66
77		44.0	85.5	68.0	50.0		88.0	48.0	124	141		27	72.5	69.5	34.0		73.0	96.5	44.5	25.0	
76		42.5		67.5	49.0		87.0	47.0	122	139		26	72.0	69.0	33.0		72.5		43.5	24.5	65
75	99.5	41.0	85.0	67.0	48.5		86.0	46.5	120	137		25	71.0		32.5		72.0	96.0	42.5		64
74	99.0	39.0		66.0	47.5		85.0	46.0	118	135		24	70.5	68.5	32.0		71.0	95.5	41.5	24.0	
73	98.5	37.5	84.5	65.5	46.5		84.5	45.5	116	132		23	70.0	68.0	31.0		70.5		41.0	23.5	63
72	98.0	36.0	84.0	65.0	45.5		83.5	45.0	114	130		22	69.5		30.5		70.0	95.0	40.0	23.0	
71	97.5	34.5		64.0	44.5	100	82.5	44.5	112	127		21	69.0	67.5	29.5		69.5	94.5	39.0	22.5	62
70	97.0	32.5	83.5	63.5	43.5	99.5	81.5	44.0	110	125		20	68.5		29.0		68.5		38.0	22.0	
69	96.0	31.0	83.0	62.5	42.5	99.0	81.0	43.5	109	123		19	68.0	67.0	28.5		68.0	94.0	37.5	21.5	61
68	95.5	29.5		62.0	41.5	98.0	80.0	43.0	107	121		18	67.0	66.5	27.5		67.5	93.5	36.5		
67	95.0	28.0	82.5	61.5	40.5	97.5	79.0	42.5	106	119		17	66.5		27.0		67.0	93.0	35.5	21.0	60
66	94.5	26.5	82.0	60.5	39.5	97.0	78.0	42.0	104	117		16	66.0	66.0	26.0		66.5		35.0	20.5	
65	94.0	25.0		60.0	38.5	96.0	77.7		102	116		15	65.5	65.5	25.5		65.5	92.5	34.0	20.0	59
64	93.5	23.5	81.5	59.5	37.5	95.5	76.5	41.5	101	114		14	65.0		25.0		65.0	92.0	33.0		
63	93.0	22.0	81.0	58.5	36.5	95.0	75.5	41.0	99	112		13	64.5	65.0	24.0		64.5		32.0		58
62	92.0	20.5		58.0	35.5	94.5	74.5	40.5	98	110		12	64.0	64.5	23.5		64.0	91.5	31.5		
61	91.5	19.0	80.5	57.0	34.5	93.5	74.0	40.0	96	108		11	63.5		23.0		63.5	91.0	30.5		
60	91.0	17.5		56.5	33.5	93.0	73.0	39.5	95	107		10	63.0	64.0	22.0		62.5	90.5	29.5		57
59	90.5	16.0	80.0	56.0	32.0	92.5	72.0	39.0	94	106		9	62.0		21.5		62.0		29.0		
58	90.0	14.5	79.5	55.0	31.0	92.0	71.0	38.5	92	104		8	61.5	63.5	20.5		61.5	90.0	28.0		
57	89.5	13.0		54.5	30.0	91.0	70.5	38.0	91	103		7	61.0	63.0	20.0		61.0	89.5	27.0		56
56	89.0	11.5	79.0	54.0	29.0	90.5	69.5		90	101		6	60.5		19.5		60.5		26.0		
55	88.0	10.0	78.5	53.0	28.0	90.0	68.5	37.5	89	100		5	60.0	62.5	18.5		60.0	89.0	25.5		55
54	87.5	8.5		52.5	27.0	89.5	68.0	37.0	87			4	59.5	62.0	18.0		59.0	88.5	24.5		
53	87.0	7.0	78.0	51.5	26.0	89.0	67.0	36.5	86			3	59.0		17.0		58.5	88.0	23.5		
52	86.5	5.5	77.5	51.0	25.0	88.0	66.0	36.0	85			2	58.0	61.5	16.5		58.0		23.0		54
51	86.0	4.0		50.5	24.0	87.5	65.0	35.5	84			1	57.5	61.0	16.0		57.5	87.5	22.0		
50	85.5	2.5	77.0	49.5	23.0	87.0	64.5	35.0	83			0	57.0		15.0		57.0	87.0	21.0		53

** The 15-T, 30-T, 45-T, 15-N, 30-N, and 45-N values are in scales of the "Rockwell" Superficial Hardness Tester, a specialized form of "Rockwell" Tester, having lighter loads and more sensitive depth reading system, used where for one or another reason the indentation must be exceptionally shallow.

Wilson Chart 38 (*continued*) *For hardened steel and alloys*

C. "ROCKWELL" HARDNESS TESTER C scale "BRALE" penetrator Load	A. "ROCKWELL" HARDNESS TESTER A scale "BRALE" penetrator—60 kg Load	D. "ROCKWELL" HARDNESS TESTER D scale "BRALE" penetrator—100 kg Load	15-N ** "ROCKWELL" SUPERFICIAL 15-N scale "N BRALE" penetrator—15 kg Load	30-N ** "ROCKWELL" SUPERFICIAL 30-N scale "N BRALE" penetrator—30 kg Load	45-N ** "ROCKWELL" SUPERFICIAL 45-N scale "N BRALE" penetrator—45 kg Load	Diamond Pyramid 4 sided 136° Diamond Pyramid 10 kg Load—Measurement of 2 diagonals by microscope. Hardness numbers computed by formula.	Bri-nell Brinell, Standard Type, 3000 kg Load Hultgren 10 mm Ball penetrator	G. "ROCKWELL" HARDNESS TESTER G scale 1/16" Ball penetrator—150 kg Load	Tensile Strength Thousand lbs per sq in.
80	92.0	86.5	96.5	92.0	87.0	1865			
79	91.5	85.5		91.5	86.5	1787			
78	91.0	84.5	96.0	91.0	85.5	1710			
77	90.5	84.0		90.5	84.5	1633			
76	90.0	83.0	95.5	90.0	83.5	1556			
75	89.5	82.5		89.0	82.5	1478			
74	89.0	81.5	95.0	88.5	81.5	1400			
73	88.5	81.0		88.0	80.5	1323			
72	88.0	80.0	94.5	87.0	79.5	1245			
71	87.0	79.5		86.5	78.5	1160			
70	86.5	78.5	94.0	86.0	77.5	1076			
69	86.0	78.0	93.5	85.0	76.5	1004			
68	85.5	77.0		84.5	75.5	942			
67	85.0	76.0	93.0	83.5	74.5	894			
66	84.5	75.5	92.5	83.0	73.0	854			
65	84.0	74.5	92.0	82.0	72.0	820			
64	83.5	74.0		81.0	71.0	789			
63	83.0	73.0	91.5	80.0	70.0	763			
62	82.5	72.5	91.0	79.0	69.0	739			
61	81.5	71.5	90.5	78.5	67.5	716			
60	81.0	71.0	90.0	77.5	66.5	695	614		314
59	80.5	70.0	89.5	76.5	65.5	675	600		306
58	80.0	69.0		75.5	64.0	655	587		299
57	79.5	68.5	89.0	75.0	63.0	636	573		291
56	79.0	67.5	88.5	74.0	62.0	617	560		284
55	78.5	67.0	88.0	73.0	61.0	598	547		277
54	78.0	66.0	87.5	72.0	59.5	580	534		270
53	77.5	65.5	87.0	71.0	58.5	562	522		263
52	77.0	64.5	86.5	70.5	57.5	545	509		256
51	76.5	64.0	86.0	69.5	56.0	528	496		250

Note:—General Conversion Values on material harder than about C-70 on the "Rockwell" Hardness Tester will probably be long in dispute because as yet there are few materials hard enough to be brought into that study and because some of those few do not flow nicely under the testing load but have a tendency to crumble and others do not approach homogeneity.

INEXACT AND ONLY FOR STEEL

C. "ROCKWELL" HARDNESS TESTER C scale "BRALE" penetrator Load	A. "ROCKWELL" HARDNESS TESTER A scale "BRALE" penetrator—60 kg Load	D. "ROCKWELL" HARDNESS TESTER D scale "BRALE" penetrator—100 kg Load	15-N ** "ROCKWELL" SUPERFICIAL 15-N scale "N BRALE" penetrator—15 kg Load	30-N ** "ROCKWELL" SUPERFICIAL 30-N scale "N BRALE" penetrator—30 kg Load	45-N ** "ROCKWELL" SUPERFICIAL 45-N scale "N BRALE" penetrator—45 kg Load	Diamond Pyramid 4 sided 136° Diamond Pyramid 10 kg Load—Measurement of 2 diagonals by microscope. Hardness numbers computed by formula.	Bri-nell Brinell, Standard Type, 3000 kg Load Hultgren 10 mm Ball penetrator	G. "ROCKWELL" HARDNESS TESTER G scale 1/16" Ball penetrator—150 kg Load	Tensile Strength Thousand lbs per sq in.
50	76.0	63.0	85.5	68.5	55.0	513	484		243
49	75.5	62.0	85.0	67.5	54.0	498	472		236
48	74.5	61.5	84.5	66.5	52.5	485	460		230
47	74.0	60.5	84.0	66.0	51.5	471	448		223
46	73.5	60.0	83.5	65.0	50.0	458	437		217
45	73.0	59.0	83.0	64.0	49.0	446	426		211
44	72.5	58.5	82.5	63.0	48.0	435	415		205
43	72.0	57.5	82.0	62.0	46.5	424	404		199
42	71.5	57.0	81.5	61.5	45.5	413	393		194
41	71.0	56.0	81.0	60.5	44.5	403	382		188
40	70.5	55.5	80.5	59.5	43.0	393	372		182
39	70.0	54.5	80.0	58.5	42.0	383	362		177
38	69.5	54.0	79.5	57.5	41.0	373	352		171
37	69.0	53.0	79.0	56.5	39.5	363	342		166
36	68.5	52.5	78.5	56.0	38.5	353	332		162
35	68.0	51.5	78.0	55.0	37.0	343	322		157
34	67.5	50.5	77.0	54.0	36.0	334	313		153
33	67.0	50.0	76.5	53.0	35.0	325	305		148
32	66.5	49.0	76.0	52.0	33.5	317	297		144
31	66.0	48.5	75.5	51.5	32.5	309	290		140
30	65.5	47.5	75.0	50.5	31.5	301	283	92.0	136
29	65.0	47.0	74.5	49.5	30.0	293	276	91.0	132
28	64.5	46.0	74.0	48.5	29.0	285	270	90.0	129
27	64.0	45.5	73.5	47.5	28.0	278	265	89.0	126
26	63.5	44.5	72.5	47.0	26.5	271	260	88.0	123
25	63.0	44.0	72.0	46.0	25.5	264	255	87.0	120
24	62.5	43.0	71.5	45.0	24.0	257	250	86.0	117
23	62.0	42.5	71.0	44.0	23.0	251	245	84.5	115
22	61.5	41.5	70.5	43.0	22.0	246	240	83.5	112
21	61.0	41.0	70.0	42.5	20.5	241	235	82.5	110
20	60.5	40.0	69.5	41.5	19.5	236	230	81.0	108

Thread Forms

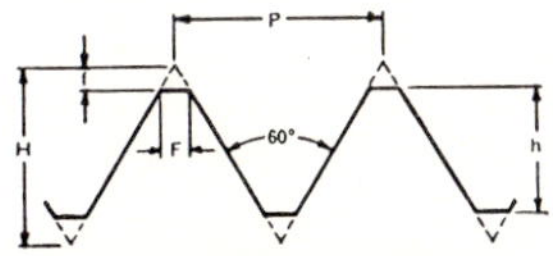

AMERICAN NATIONAL
h =.649519xP = DEPTH OF THREAD
H =.866025xP F =.125xP
f =.108253xP

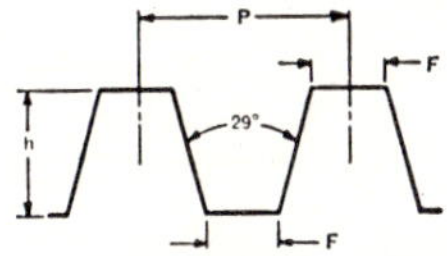

ACME
h =.500xP = DEPTH OF THREAD
F =.3707xP

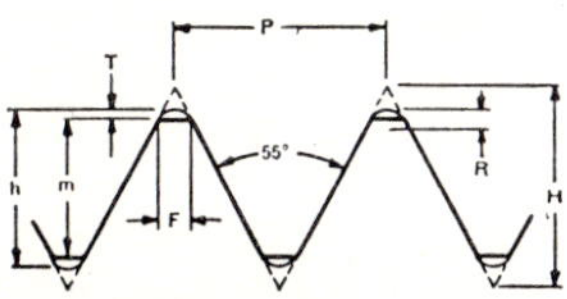

WHITWORTH MOD. & FULL FORM
BRITISH STRAIGHT PIPE
m =.4925xP = DEPTH OF MOD. TH'D.
h =.64033xP = DEPTH OF F.F. TH'D.
R =.1373xP = RADIUS
T =.0739xP
H =.96049xP F =.24357xP

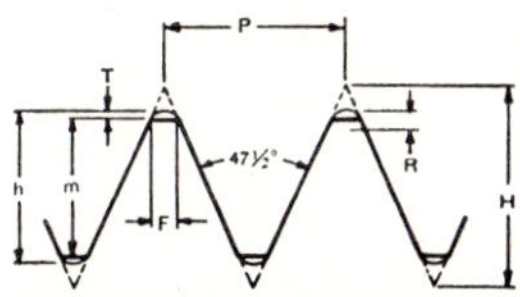

BRITISH ASSOCIATION
m =.3825xP = DEPTH OF MOD. TH'D.
h =.6xP = DEPTH OF F. F. THREAD
F =.6634xP
R =.1818xP = RADIUS
T =.1087xP H =1.13634xP

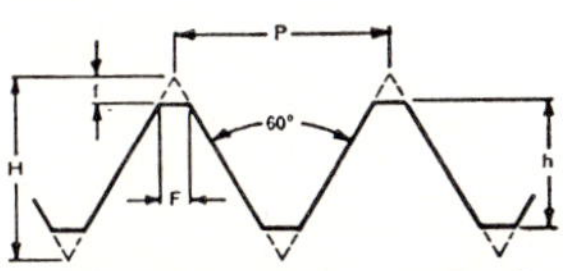

INTERNATIONAL METRIC
h =.649519xP = DEPTH OF THREAD
H =.866025xP F =.125xP
f =.108253xP

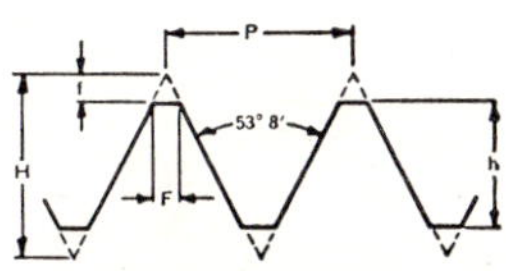

LOEWENHERZ
h =.75xP = DEPTH OF THREAD
H =1.00xP F =.125xP
f =.125xP

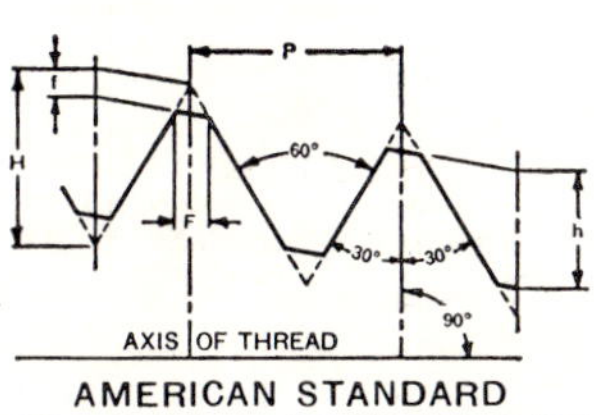

AMERICAN STANDARD
TAPER PIPE GAGE FORM N.P.T.
H =.866025xP h =.666025xP
f =.1xP F =.11547xP

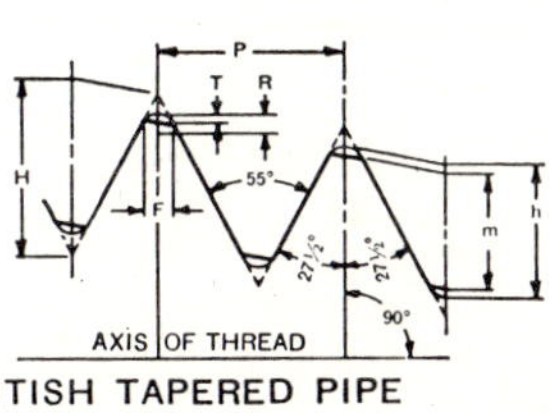

BRITISH TAPERED PIPE
m = .4925xP = DEPTH OF MOD. THREAD
h = .64033xP = DEPTH OF FULL FORM TH'D.
H = .96024xP T = .0739xP
R = .1373xP = RADIUS F = .24357xP

Screw Thread Standards For Federal Services

Go and Not Go Pitch Diameter Limits, Classes 1 — 2 — 3 — 4
National Coarse (NC) and National Fine (NF) Series

All Dimensions Given in Inches

Size	SCREWS (Ring Thread Gages)				Basic P. D.	NUTS (Plug Thread Gages)			
	Loose Fit Class 1	Free Fit Class 2	Med. Fit Class 3	Close Fit Class 4		Loose Fit Class 1	Free Fit Class 2	Med. Fit Class 3	Close Fit Class 4
0–80NF	.0512 .0488	.0519 .0502	.0519 .0506		.0519	.0519 .0543	.0519 .0536	.0519 .0532	
1–72NF	.0633 .0608	.0640 .0622	.0640 .0627		.0640	.0640 .0665	.0640 .0658	.0640 .0653	
1–64NC	.0622 .0596	.0629 .0610	.0629 .0615		.0629	.0629 .0655	.0629 .0648	.0629 .0643	
2–64NF	.0752 .0726	.0759 .0740	.0759 .0745		.0759	.0759 .0785	.0759 .0778	.0759 .0773	
2–56NC	.0736 .0708	.0744 .0724	.0744 .0729		.0744	.0744 .0772	.0744 .0764	.0744 .0759	
3–56NF	.0866 .0838	.0874 .0854	.0874 .0859		.0874	.0874 .0902	.0874 .0894	.0874 .0889	
3–48NC	.0846 .0815	.0855 .0833	.0855 .0839		0855	.0855 .0886	.0855 .0877	.0855 .0871	
4–48NF	.0976 .0945	.0985 .0963	.0985 .0969		.0985	.0985 .1016	.0985 .1007	.0985 .1001	
4–40NC	.0948 .0914	.0958 .0934	.0958 .0941		.0958	.0958 .0992	.0958 .0982	.0958 .0975	
5–44NF	.1093 .1061	.1102 .1079	.1102 .1086		.1102	.1102 .1134	.1102 .1125	.1102 .1118	
5–40NC	.1078 .1044	.1088 .1064	.1088 .1071		.1088	.1088 .1122	.1088 .1112	.1088 .1105	
6–40NF	.1208 .1174	.1218 .1194	.1218 .1201		.1218	.1218 .1252	.1218 .1242	.1218 .1235	
6–32NC	.1166 .1128	.1177 .1150	.1177 .1158		.1177	.1177 .1215	.1177 .1204	.1177 .1196	
8–36NF	.1449 .1413	.1460 .1435	.1460 .1442		.1460	.1460 .1496	.1460 .1485	.1460 .1478	
8–32NC	.1426 .1388	.1437 .1410	.1437 .1418		.1437	.1437 .1475	.1437 .1464	.1437 .1456	
10–32NF	.1686 .1648	.1697 .1670	.1697 .1678		.1697	.1697 .1735	.1697 .1724	.1697 .1716	
10–24NC	.1616 .1570	.1629 .1596	.1629 .1605		.1629	.1629 .1675	.1629 .1662	.1629 .1653	
12–28NF	.1916 .1873	.1928 .1897	.1928 .1906		.1928	.1928 .1971	.1928 .1959	.1928 .1950	
12–24NC	.1876 .1830	.1889 .1856	.1889 .1865		.1889	.1889 .1935	.1889 .1922	.1889 .1913	
¼–28NF	.2256 .2213	.2268 .2237	.2268 .2246	.2270 .2259	.2268	.2268 .2311	.2268 .2299	.2268 .2290	.2268 .2279
¼–20NC	.2160 .2109	.2175 .2139	.2175 .2149	.2178 .2165	.2175	.2175 .2226	.2175 .2211	.2175 .2201	.2175 .2188
5/16–24NF	.2841 .2795	.2854 .2821	.2854 .2830	.2857 .2845	.2854	.2854 .2900	.2854 .2887	.2854 .2878	.2854 .2866
5/16–18NC	.2748 .2691	.2764 .2723	.2764 .2734	.2767 .2752	.2764	.2764 .2821	.2764 .2805	.2764 .2794	.2764 .2779

Screw Thread Standards For Federal Services

All Dimensions Given in Inches

Size	SCREWS (Ring Thread Gages)				Basic P. D.	NUTS (Plug Thread Gages)			
	Loose Fit Class 1	Free Fit Class 2	Med. Fit Class 3	Close Fit Class 4		Loose Fit Class 1	Free Fit Class 2	Med. Fit Class 3	Close Fit Class 4
3/8-24NF	.3466 .3420	.3479 .3446	.3479 .3455	.3482 .3470	.3479	.3479 .3525	.3479 .3512	.3479 .3503	.3479 .3491
3/8-16NC	.3326 .3263	.3344 .3299	.3344 .3312	.3348 .3332	.3344	.3344 .3407	.3344 .3389	.3344 .3376	.3344 .3360
7/16-20NF	.4035 .3984	.4050 .4014	.4050 .4024	.4053 .4040	.4050	.4050 .4101	.4050 .4086	.4050 .4076	.4050 .4063
7/16-14NC	.3890 .3820	.3911 .3862	.3911 .3875	.3915 .3897	.3911	.3911 .3981	.3911 .3960	.3911 .3947	.3911 .3929
1/2-20NF	.4660 .4609	.4675 .4639	.4675 .4649	.4678 .4665	.4675	.4675 .4726	.4675 .4711	.4675 .4701	.4675 .4688
1/2-13NC	.4478 .4404	.4500 .4448	.4500 .4463	.4504 .4485	.4500	.4500 .4574	.4500 .4552	.4500 .4537	.4500 .4519
9/16-18NF	.5248 .5191	.5264 .5223	.5264 .5234	.5267 .5252	.5264	.5264 .5321	.5264 .5305	.5264 .5294	.5264 .5279
9/16-12NC	.5060 .4981	.5084 .5028	.5084 .5044	.5089 .5069	.5084	.5084 .5163	.5084 .5140	.5084 .5124	.5084 .5104
5/8-18NF	.5873 .5816	.5889 .5848	.5889 .5859	.5892 .5877	.5889	.5889 .5946	.5889 .5930	.5889 .5919	.5889 .5904
5/8-11NC	.5634 .5549	.5660 .5601	.5660 .5618	.5665 .5644	.5660	.5660 .5745	.5660 .5719	.5660 .5702	.5660 .5681
3/4-16NF	.7076 .7013	.7094 .7049	.7094 .7062	.7098 .7082	.7094	.7094 .7157	.7094 .7139	.7094 .7126	.7094 .7110
3/4-10NC	.6822 .6730	.6850 .6786	.6850 .6805	.6856 .6833	.6850	.6850 .6942	.6850 .6914	.6850 .6895	.6850 .6873
7/8-14NF	.8265 .8195	.8286 .8237	.8286 .8250	.8290 .8272	.8286	.8286 .8356	.8286 .8335	.8286 .8322	.8286 .8304
7/8- 9NC	.7997 .7897	.8028 .7958	.8028 .7979	.8034 .8010	.8028	.8028 .8128	.8028 .8098	.8028 .8077	.8028 .8052
1-14NF	.9515 .9445	.9536 .9487	.9536 .9500	.9540 .9522	.9536	.9536 .9606	.9536 .9585	.9536 .9572	.9536 .9554
1- 8NC	.9154 .9043	.9188 .9112	.9188 .9134	.9195 .9168	.9188	.9188 .9299	.9188 .9264	.9188 .9242	.9188 .9215
1 1/8-12NF	1.0685 1.0606	1.0709 1.0653	1.0709 1.0669	1.0714 1.0694	1.0709	1.0709 1.0788	1.0709 1.0765	1.0709 1.0749	1.0709 1.0729
1 1/8- 7NC	1.0283 1.0159	1.0322 1.0237	1.0322 1.0263	1.0330 1.0300	1.0322	1.0322 1.0446	1.0322 1.0407	1.0322 1.0381	1.0322 1.0352
1 1/4-12NF	1.1935 1.1856	1.1959 1.1903	1.1959 1.1919	1.1964 1.1944	1.1959	1.1959 1.2038	1.1959 1.2015	1.1959 1.1999	1.1959 1.1979
1 1/4- 7NC	1.1533 1.1409	1.1572 1.1487	1.1572 1.1513	1.1580 1.1550	1.1572	1.1572 1.1696	1.1572 1.1657	1.1572 1.1631	1.1572 1.1602
1 3/8-12NF	1.3185 1.3106	1.3209 1.3153	1.3209 1.3169	1.3214 1.3194	1.3209	1.3209 1.3288	1.3209 1.3265	1.3209 1.3249	1.3209 1.3229
1 3/8- 6NC	1.2623 1.2478	1.2667 1.2566	1.2667 1.2596	1.2676 1.2640	1.2667	1.2667 1.2812	1.2667 1.2768	1.2667 1.2738	1.2667 1.2703
1 1/2-12NF	1.4435 1.4356	1.4459 1.4403	1.4459 1.4419	1.4464 1.4444	1.4459	1.4459 1.4538	1.4459 1.4515	1.4459 1.4499	1.4459 1.4479
1 1/2- 6NC	1.3873 1.3728	1.3917 1.3816	1.3917 1.3846	1.3926 1.3890	1.3917	1.3917 1.4062	1.3917 1.4018	1.3917 1.3988	1.3917 1.3953

Basic Thread Dimensions

American National Form

Nominal Size	Major Diameter Inches	Pitch Diameter Inches	Minor Diameter of Ring Gage 83 1/3% of Full Thread Inches
0-80NF	.0600	.0519	.0465
1-64NC	.0730	.0629	.0561
1-72NF	.0730	.0640	.0580
2-56NC	.0860	.0744	.0667
2-64NF	.0860	.0759	.0691
3-48NC	.0990	.0855	.0764
3-56NF	.0990	.0874	.0797
4-36NS	.1120	.0940	.0819
4-40NC	.1120	.0958	.0849
4-48NF	.1120	.0985	.0894
5-40NC	.1250	.1088	.0979
5-44NF	.1250	.1102	.1004
6-32NC	.1380	.1177	.1042
6-40NF	.1380	.1218	.1109
8-32NC	.1640	.1437	.1302
8-36NF	.1640	.1460	.1339
10-24NC	.1900	.1629	.1449
10-30NS	.1900	.1684	.1539
10-32NF	.1900	.1697	.1562
12-24NC	.2160	.1889	.1709
12-28NF	.2160	.1928	.1773
14-20NS	.2420	.2095	.1879
14-24NS	.2420	.2149	.1969
1/4-20NC	.2500	.2175	.1959
1/4-24NS	.2500	.2229	.2049
1/4-28NF	.2500	.2268	.2113
1/4-32NS	.2500	.2297	.2162
5/16-18NC	.3125	.2764	.2524
5/16-24NF	.3125	.2854	.2674
5/16-32NS	.3125	.2922	.2787
3/8-16NC	.3750	.3344	.3073
3/8-24NF	.3750	.3479	.3299
7/16-14NC	.4375	.3911	.3602
7/16-20NF	.4375	.4050	.3834
1/2-13NC	.5000	.4500	.4167
1/2-20NF	.5000	.4675	.4459
9/16-12NC	.5625	.5084	.4723
9/16-18NF	.5625	.5264	.5024
5/8-11NC	.6250	.5660	.5266
5/8-18NF	.6250	.5889	.5649
11/16-11NS	.6875	.6285	.5891
11/16-16NS	.6875	.6469	.6198
3/4-10NC	.7500	.6850	.6417
3/4-16NF	.7500	.7094	.6823
7/8- 9NC	.8750	.8028	.7547
7/8-14NF	.8750	.8286	.7977
7/8-18NS	.8750	.8389	.8149
1 - 8NC	1.0000	.9188	.8647
1 -14NF	1.0000	.9536	.9227
1 1/8- 7NC	1.1250	1.0322	.9704
1 1/8-12NF	1.1250	1.0709	1.0348
1 1/4- 7NC	1.2500	1.1572	1.0954
1 1/4-12NF	1.2500	1.1959	1.1598
1 3/8- 6NC	1.3750	1.2667	1.1946
1 3/8-12NF	1.3750	1.3209	1.2848
1 1/2- 6NC	1.5000	1.3917	1.3196
1 1/2-12NF	1.5000	1.4459	1.4098

Screw Thread Standards For Federal Service

Go and Not Go Pitch Diameter Limits

Classes 2 and 3 National Extra Fine (NEF) Series

Size	Screws (Ring Thread Gages)		Nuts (Plug Thread Gages)	
	Class 2	Class 3	Class 2	Class 3
¼–32	.2297 .2265	.2297 .2275	.2297 .2329	.2297 .2319
5/16–32	.2922 .2889	.2922 .2899	.2922 .2955	.2922 .2945
⅜–32	.3547 .3513	.3547 .3523	.3547 .3581	.3547 .3571
7/16–28	.4143 .4107	.4143 .4118	.4143 .4179	.4143 .4168
½–28	.4768 .4731	.4768 .4742	.4768 .4805	.4768 .4794
9/16–24	.5354 .5314	.5354 .5326	.5354 .5394	.5354 .5382
⅝–24	.5979 .5938	.5979 .5950	.5979 .6020	.5979 .6008
11/16–24	.6604 .6563	.6604 .6575	.6604 .6645	.6604 .6633
¾–20	.7175 .7129	.7175 .7143	.7175 .7221	.7175 .7207
13/16–20	.7800 .7754	.7800 .7768	.7800 .7846	.7800 .7832
⅞–20	.8425 .8378	.8425 .8392	.8425 .8472	.8425 .8458
15/16–20	.9050 .9003	.9050 .9017	.9050 .9097	.9050 .9083
1 –20	.9675 .9627	.9675 .9641	.9675 .9723	.9675 .9709
1 1/16–18	1.0264 1.0216	1.0264 1.0228	1.0264 1.0312	1.0264 1.0300
1 ⅛–18	1.0889 1.0837	1.0889 1.0853	1.0889 1.0941	1.0889 1.0925
1 3/16–18	1.1514 1.1462	1.1514 1.1478	1.1514 1.1566	1.1514 1.1550
1 ¼–18	1.2139 1.2086	1.2139 1.2102	1.2139 1.2192	1.2139 1.2176
1 5/16–18	1.2764 1.2711	1.2764 1.2727	1.2764 1.2817	1.2764 1.2801
1 ⅜–18	1.3389 1.3335	1.3389 1.3351	1.3389 1.3443	1.3389 1.3427
1 7/16–18	1.4014 1.3960	1.4014 1.3976	1.4014 1.4068	1.4014 1.4052
1 ½–18	1.4639 1.4584	1.4639 1.4601	1.4639 1.4694	1.4639 1.4677
1 9/16–18	1.5264 1.5209	1.5264 1.5225	1.5264 1.5319	1.5264 1.5303
1 ⅝–18	1.5889 1.5833	1.5889 1.5850	1.5889 1.5945	1.5889 1.5928
1 11/16–18	1.6514 1.6458	1.6514 1.6475	1.6514 1.6570	1.6514 1.6553
1 ¾–16	1.7094 1.7035	1.7094 1.7053	1.7094 1.7153	1.7094 1.7135
2 –16	1.9594 1.9533	1.9594 1.9551	1.9594 1.9655	1.9594 1.9637

Basic Thread Dimensions

Whitworth Standard Threads and Modified Gage Threads
(Including British Standard Fine)

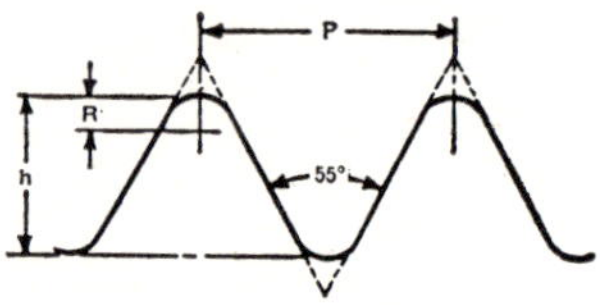

WHITWORTH STANDARD

h=.64033xP= DEPTH OF THREAD
R=.1373xP= RADIUS

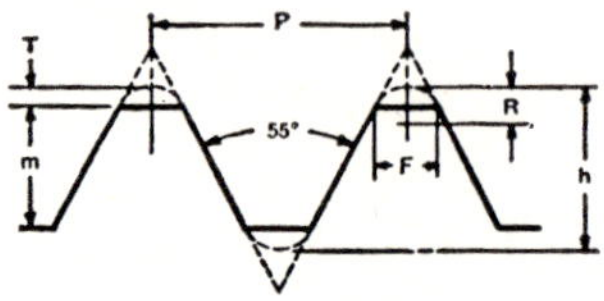

WHITWORTH MODIFIED

m=.4925xP= DEPTH OF THREAD
h=.64033xP F=.24357xP
R=.1373xP= RADIUS
T=.0739xP

Nominal Size	Major Diameter Full Form Inches	Minor Diameter Full Form Inches	Pitch Diameter Inches	Major Diameter Modified Form Inches	Minor Diameter Modified Form Inches
1/8–40	.1250	.0930	.1090	.1213	.0967
3/16–24	.1875	.1341	.1608	.1813	.1403
3/16–32	.1875	.1475	.1675	.1829	.1521
7/32–28	.2188	.1730	.1959	.2135	.1783
1/4–20	.2500	.1860	.2180	.2426	.1934
1/4–26	.2500	.2008	.2254	.2443	.2065
9/32–26	.2812	.2320	.2566	.2755	.2377
5/16–18	.3125	.2413	.2769	.3043	.2495
5/16–22	.3125	.2543	.2834	.3058	.2610
3/8–16	.3750	.2950	.3350	.3658	.3042
3/8–20	.3750	.3110	.3430	.3676	.3184
7/16–14	.4375	.3461	.3918	.4270	.3566
7/16–18	.4375	.3663	.4019	.4293	.3745
1/2–12	.5000	.3932	.4466	.4876	.4056
1/2–16	.5000	.4200	.4600	.4908	.4292
9/16–12	.5625	.4557	.5091	.5501	.4681
9/16–16	.5625	.4825	.5225	.5533	.4917
5/8–11	.6250	.5086	.5668	.6116	.5220
5/8–14	.6250	.5336	.5793	.6145	.5441
11/16–11	.6875	.5711	.6293	.6741	.5845
11/16–14	.6875	.5961	.6418	.6770	.6066

Basic Thread Dimensions

Whitworth Standard Threads and Modified Gage Threads

(Including British Standard Fine)

Nominal Size	Major Diameter Full Form Inches	Minor Diameter Full Form Inches	Pitch Diameter Inches	Major Diameter Modified Form Inches	Minor Diameter Modified Form Inches
¾–10	.7500	.6220	.6860	.7353	.6368
¾–12	.7500	.6432	.6966	.7376	6555
13/16–12	.8125	.7057	7591	.8001	.7181
⅞– 9	.8750	.7328	.8039	.8586	.7492
⅞–11	.8750	.7586	.8168	.8616	.7720
1 – 8	1 0000	.8400	.9200	.9816	.8584
1 –10	1.0000	.8720	.9360	.9853	.8868
1 ⅛– 7	1.1250	.9420	1.0335	1.1039	.9631
1 ⅛– 9	1.1250	.9828	1.0539	1.1086	.9992
1 ¼– 7	1.2500	1.0670	1.1585	1.2289	1.0881
1 ¼– 9	1.2500	1.1078	1.1789	1.2336	1.1242
1 ⅜– 8	1.3750	1.2150	1.2950	1.3566	1.2334
1 ½– 6	1.5000	1.2866	1.3933	1 4754	1.3112
1 ½– 8	1.5000	1.3400	1.4200	1.4816	1.3584
1 ⅝– 8	1.6250	1.4650	1.5450	1.6066	1.4834
1 ¾– 5	1.7500	1.4939	1.6219	1.7204	1.5234
1 ¾– 7	1.7500	1.5670	1.6585	1.7289	1.5881
2 – 4½	2.0000	1.7154	1.8577	1.9671	1.7483
2 – 7	2.0000	1.8170	1.9085	1.9789	1.8381
2 ¼– 4	2.2500	1.9298	2.0899	2.2130	1.9668
2 ¼– 6	2.2500	2.0366	2:1433	2.2254	2.0612
2 ½– 4	2.5000	2.1798	2.3399	2.4630	2.2168
2 ½– 6	2.5000	2.2866	2.3933	2.4754	2.3112
2 ¾– 3½	2.7500	2.3840	2.5670	2.7077	2.4263
2 ¾– 6	2.7500	2.5366	2.6433	2.7254	2.5612
3 – 3½	3.0000	2.6340	2.8170	2.9577	2.6763
3 – 5	3.0000	2.7438	2.8719	2.9704	2.7734
3 ¼– 3¼	3.2500	2.8560	3.0530	3.2045	2.9015
3 ¼– 5	3.2500	2.9938	3.1219	3.2204	3.0234
3 ½– 3¼	3.5000	3.1060	3.3030	3.4545	3.1515
3 ½– 4½	3.5000	3.2154	3.3577	3.4671	3.2483
3 ¾– 3	3.7500	3.3232	3.5366	3.7008	3.3724
3 ¾– 4½	3.7500	3.4654	3.6077	3.7171	3.4983
4 – 3	4.0000	3.5732	3.7866	3.9508	3.6224
4 – 4½	4.0000	3.7154	3.8577	3.9671	3.7483

Basic Thread Dimensions

British Association Standard

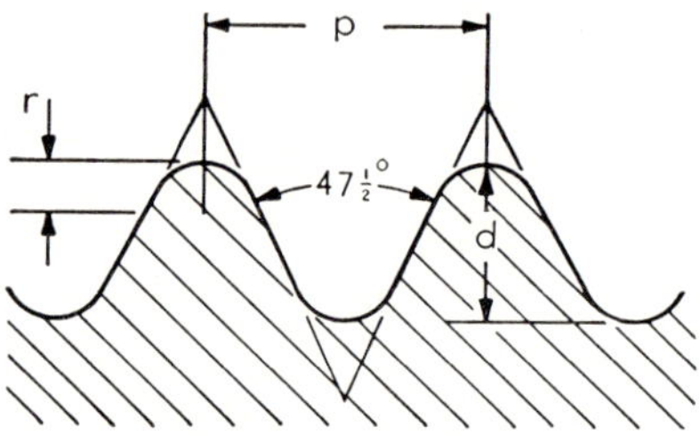

Formula $\begin{cases} p = \text{pitch} \\ d = \text{depth} = p \times .6 \\ r = \text{radius} = \dfrac{2 \times p}{11} \end{cases}$

Number	Pitch m/m	Major Diameter m/m	Pitch Diameter m/m	Root Diameter m/m
0	1.00	6.0	5.400	4.80
1	.90	5.3	4.760	4.22
2	.81	4.7	4.215	3.73
3	.73	4.1	3.660	3.22
4	.66	3.6	3.205	2.81
5	.59	3.2	2.845	2.49
6	.53	2.8	2.480	2.16
7	.48	2.5	2.210	1.92
8	.43	2.2	1.940	1.68
9	.39	1.9	1.665	1.43
10	.35	1.7	1.490	1.28
11	.31	1.5	1.315	1.13
12	.28	1.3	1.130	.96
14	.23	1.0	.860	.72

Basic Thread Dimensions

Acme Threads

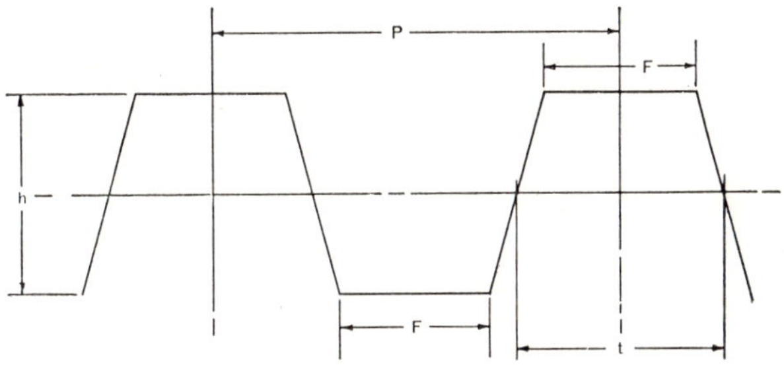

h = .5P. = Depth of Thread
t = .5P. F = .3707P.

All Dimensions Given in Inches

Threads per Inch	Pitch P	Depth of Thread h	Thread Thickness t	Flat F
16	0.06250	0.03125	0.03125	0.0232
14	.07143	.03571	.03571	.0265
12	.08333	.04167	.04167	.0309
10	.10000	.05000	.05000	.0371
8	.12500	.06250	.06250	.0463
6	.16667	.08333	.08333	.0618
5	.20000	.10000	.10000	.0741
4	.25000	.12500	.12500	.0927
3	.33333	.16667	.16667	.1236
2½	.40000	.20000	.20000	.1483
2	.50000	.25000	.25000	.1853
1½	.66667	.33333	.33333	.2471
1⅓	.75000	.37500	.37500	.2780
1	1.00000	.50000	.50000	.3707

Recommended Wire Size for Checking 29° Acme Threads

All Dimensions Given in Inches

Threads per Inch	Best Wire Size
1	.51645
1⅓	.38734
1½	.34430
2	.25822
2½	.20658
3	.17215
4	.12911
5	.10329
6	.08608
8	.06456
10	.05164
12	.04304
14	.03689
16	.03228

American Standard Pipe Threads and Pipe Gage Threads

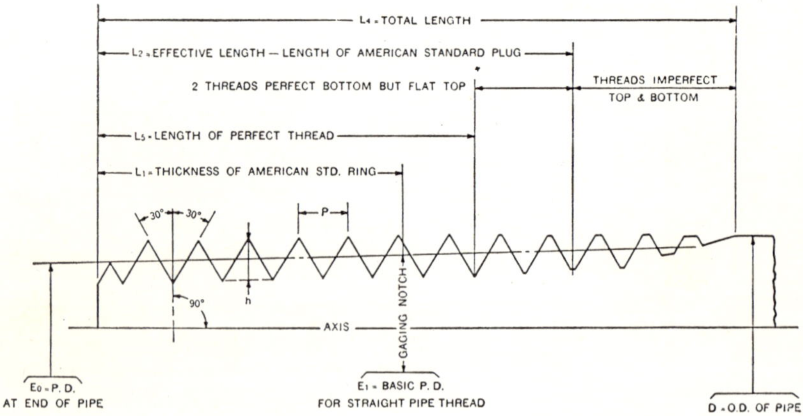

N = NUMBER OF THREADS PER INCH
¾ INCH TAPER PER FOOT ON DIAMETER

h = DEPTH OF PIPE THREADS = .8P
h = DEPTH OF GAGE THREADS = .666025P

American Standard Pipe Threads and Pipe Gage Threads

with Locknut Threads and Basic Straight Pipe Sizes

Pipe Size Inches	Outside Diameter Pipe Inches	Threads per Inch	Pitch Inches	Depth of Thread .8 P. Inches	Pitch Diameter at Gaging Notch Basic Straight Inches	Pitch Diameter at end of Pipe Inches	Thickness of American St'd. Ring Gage (also Distance from small end to gaging notch) Inches	Length of American St'd. Plug (Effective Length of Thread on Pipe) Inches	Length of Perfect Thread Inches	Total Length of Thread Inches
	D	N	P	h	E_1	E_0	L_1	L_2	L_5	L_4
1/16	.3125	27	.03704	.0296	.2812	.2712	.160	.2611	.1870	.3896
1/8	.405	27	.03704	.0296	.3748	.3635	.180	.2639	.1898	.3924
1/4	.540	18	.05556	.0444	.4899	.4774	.200	.4018	.2907	.5946
3/8	.675	18	.05556	.0444	.6270	.6120	.240	.4078	.2967	.6006
1/2	.840	14	.07143	.0571	.7784	.7584	.320	.5337	.3909	.7815
3/4	1.050	14	.07143	.0571	.9889	.9677	.339	.5457	.4029	.7935
1	1.315	11 1/2	.08696	.0696	1.2386	1.2136	.400	.6828	.5088	.9845
1 1/4	1.660	11 1/2	.08696	.0696	1.5834	1.5571	.420	.7068	.5329	1.0085
1 1/2	1.900	11 1/2	.08696	.0696	1.8223	1.7961	.420	.7235	.5496	1.0252
2	2.375	11 1/2	.08696	.0696	2.2963	2.2690	.436	.7565	.5826	1.0582
2 1/2	2.875	8	.12500	.1000	2.7622	2.7195	.682	1.1375	.8875	1.5713
3	3.500	8	.12500	.1000	3.3885	3.3406	.766	1.2000	.9500	1.6337
3 1/2	4.000	8	.12500	.1000	3.8888	3.8375	.821	1.2500	1.0000	1.6837
4	4.500	8	.12500	.1000	4.3871	4.3344	.844	1.3000	1.0500	1.7337
5	5.563	8	.12500	.1000	5.4493	5.3907	.937	1.4063	1.1563	1.8407
6	6.625	8	.12500	.1000	6.5060	6.4461	.958	1.5125	1.2625	1.9467
8	8.625	8	.12500	.1000	8.5000	8.4336	1.063	1.7125	1.4625	2.1467
10	10.750	8	.12500	.1000	10.6209	10.5453	1.210	1.9250	1.6750	2.3587
12	12.750	8	.12500	.1000	12.6178	12.5328	1.360	2.1250	1.8750	2.5587

British Standard Pipe Threads and Pipe Gage Threads

(American Practice)

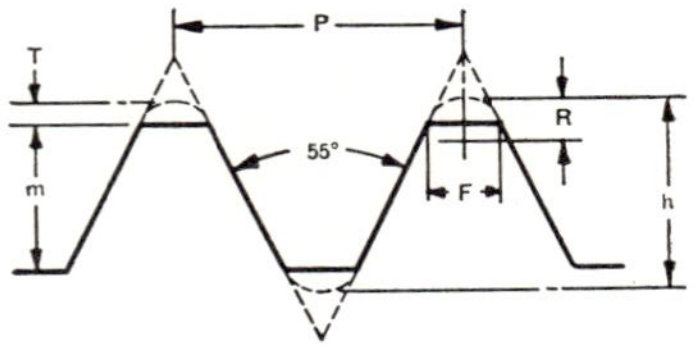

WHITWORTH MODIFIED THREAD

m =.4925 X P = DEPTH OF THREAD
h =.64033 X P   R=.1373 X P = RADIUS
F =.24357 X P   T =.0739 X P
N = NUMBER OF THREADS PER INCH
3/4 TAPER PER FOOT ON DIAMETER

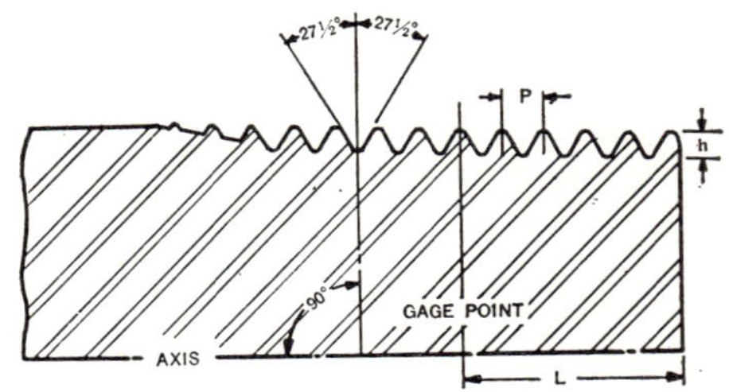

Product accepted by gages made to American Practice (which is the same practice as is used in gaging American National taper pipe threads) will interchange with the product checked by the gages to the British Practice as outlined by the British Engineering Standards Association.

British Standard Pipe Threads and Pipe Gage Threads (Taper)

(American Practice)

All Dimensions Given in Inches

Pipe Size Nominal Inside Diameter	Approx. Outside Diameter of Pipe	Threads per Inch	Pitch	Depth of Full Form Pipe Thread	Truncation from Single Thread Depth on Gage Threads	Pitch Diameter at Gaging Point	Distance from Small End to Gaging Point Gage Length	Major Diameter of Pipe Thread at Gaging Point	Minor Diameter of Pipe Thread at Gaging Point	Major Dia. of Plug Gage at Gaging Point P&W Modified Form	Minor Dia. of Ring Gage at Gaging Point P&W Modified Form
		N	P	h	T		L				
1/8	13/32	28	.03571	.0229	.0026	.3601	.1563	.383	.3372	.3778	.3424
1/4	17/32	19	.05263	.0337	.0039	.4843	.2367	.518	.4506	.5102	.4584
3/8	11/16	19	.05263	.0337	.0039	.6223	.2500	.656	.5886	.6482	.5964
1/2	27/32	14	.07143	.0457	.0053	.7793	.3214	.825	.7336	.8144	.7442
3/4	1 1/16	14	.07143	.0457	.0053	.9953	.3750	1.041	.9496	1:0304	.9602
1	1 11/32	11	.09091	.0582	.0067	1.2508	.4091	1.309	1.1926	1.2956	1.2060
1 1/4	1 11/16	11	.09091	.0582	.0067	1.5918	.5000	1.650	1.5336	1.6366	1.5470
1 1/2	1 29/32	11	.09091	.0582	.0067	1.8238	.5000	1.882	1.7656	1.8686	1.7790
2	2 3/8	11	.09091	.0582	.0067	2.2888	.6250	2.347	2.2306	2.3336	2.2440
2 1/2	3	11	.09091	.0582	.0067	2.9018	.6875	2.960	2.8436	2.9466	2.8570
3	3 1/2	11	.09091	.0582	.0067	3.4018	.8125	3.460	3.3436	3.4466	3.3570
3 1/2	4	11	.09091	.0582	.0067	3.8918	.8750	3.950	3.8336	3.9366	3.8470
4	4 1/2	11	.09091	.0582	.0067	4.3918	1.0000	4.450	4.3336	4.4366	4.3470
5	5 1/2	11	.09091	.0582	.0067	5.3918	1.1250	5.450	5.3336	5.4366	5.3470
6	6 1/2	11	.09091	.0582	.0067	6.3918	1.1250	6.450	6.3336	6.4366	6.3470

American Standard Machine Tapers

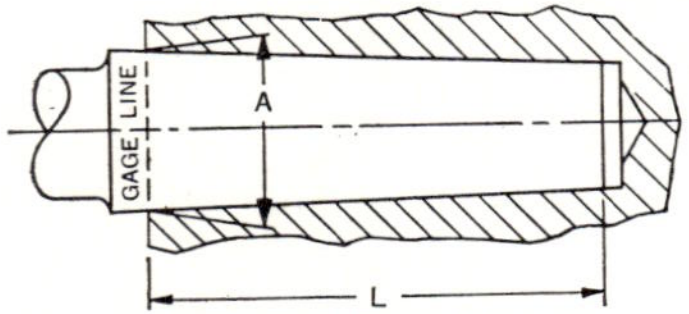

Self Holding Tapers

A.S.A. Dimensions				
No. of Taper	Taper Inches per Foot	Diameter at End of Socket Inches A	Standard Plug Depth Inches L	Origin of Series
.239	.50200	0.23922	15/16	Brown and Sharpe Taper Series
.299	.50200	.29968	1 3/16	
.375	.50200	.37525	1 1/2	
1	.59858	.47500	2 1/8	Morse Taper Series
2	.59941	.70000	2 9/16	
3	.60235	.93800	3 3/16	
4	.62326	1.23100	4 1/16	
4 1/2	.62400	1.50000	4 1/2	
5	.63151	1.74800	5 3/16	
6	.62565	2.49400	7 1/4	
200	.750	2.0000	4 3/4	3/4 Inch per Foot Taper Series
250	.750	2.5000	5 1/2	
300	.750	3.0000	6 1/4	
350	.750	3.5000	7	
400	.750	4.0000	7 3/4	
450	.750	4.5000	8 1/2	
500	.750	5.0000	9 1/4	
600	.750	6.0000	10 3/4	
800	.750	8.0000	13 3/4	
1000	.750	10.0000	16 3/4	
1200	.750	12.0000	19 3/4	

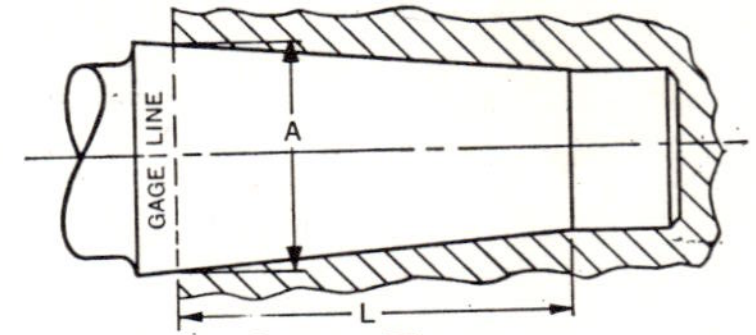

Steep Tapers

Basic Dimensions		
Number of Taper	Diameter at End of Socket Inches A	Standard Plug Depth Inches L
10	.625	.8750
20	.875	1.3125
30	1.250	1.8750
40	1.750	2.6875
50	2.750	4.0000
60	4.250	6.3750

Taper per Foot 3.500 Inches

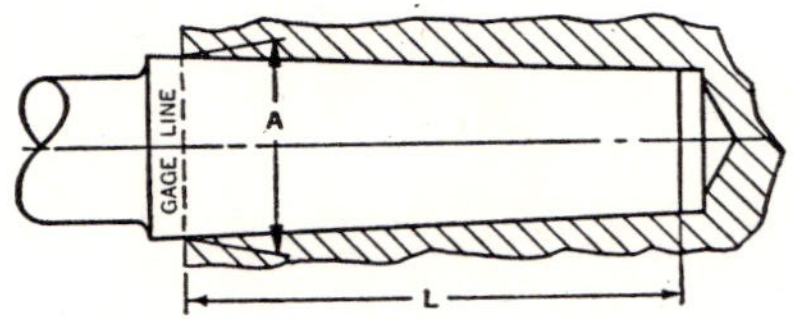

Taper Dimensions

Brown & Sharpe—Morse—Jarno

BROWN & SHARPE				MORSE				JARNO			
No. of Taper	Taper Inches per Foot	Diameter at End of Socket Inches A	Standard Plug Depth Inches L	No. of Taper	Taper Inches per Foot	Diameter at End of Socket Inches A	Standard Plug Depth Inches L	No. of Taper	Taper Inches per Foot	Diameter at End of Socket Inches A	Standard Plug Depth Inches L
1	0.50200	0.23922	15/16	0	0.62460	0.3561	2	1	0.6000	0.125	1/2
2	.50200	.29968	1 3/16	1	.59858	.475	2 1/8	2	.6000	.250	1
3	.50200	.37525	1 1/2	2	.59941	.700	2 9/16	3	.6000	.375	1 1/2
4	.50240	.42065	1 11/16	3	.60235	.938	3 3/16	4	.6000	.500	2
5	.50160	.53882	2 1/8	4	.62326	1.231	4 1/16	5	.6000	.625	2 1/2
6	.50329	.59961	2 3/8	5	.63151	1.748	5 3/16	6	.6000	.750	3
7	.50147	.72014	2 7/8	6	.62565	2.494	7 1/4	7	.6000	.875	3 1/2
8	.50100	.89873	3 9/16	7	.62400	3.270	10	8	.6000	1.000	4
9	.50085	1.07748	4 1/4					9	.6000	1.125	4 1/2
10	.51612	1.25970	5					10	.6000	1.250	5
11	.50100	1.49784	5 15/16					11	.6000	1.375	5 1/2
12	.49973	1.79681	7 1/8					12	.6000	1.500	6
13	.50020	2.07310	7 3/4					13	.6000	1.625	6 1/2
14	.50000	2.34375	8 1/4					14	.6000	1.750	7
15	.50000	2.61458	8 3/4					15	.6000	1.875	7 1/2
16	.50000	2.88542	9 1/4					16	.6000	2.000	8
17	.50000	3.15625	9 3/4					17	.6000	2.125	8 1/2
18	.50000	3.42708	10 1/4					18	.6000	2.250	9
								19	.6000	2.375	9 1/2
								20	.6000	2.500	10

Decimal Equivalents of Millimeters

(Basis: 1 inch = 25.4 millimeters)

M/M	Inches	M/M	Inches	M/M	Inches	M/M	Inches
.1	.00394	16.0	.62992	35.5	1.39764	62	2.44095
.2	.00787	16.5	.64961	36.0	1.41732	63	2.48032
.3	.01181	17.0	.66929	36.5	1.43701	64	2.51969
.4	.01575	17.5	.68898	37.0	1.45669	65	2.55906
.5	.01969	18.0	.70866	37.5	1.47638	66	2.59843
.6	.02362	18.5	.72835	38.0	1.49606	67	2.63780
.7	.02756	19.0	.74803	38.5	1.51575	68	2.67717
.8	.03150	19.5	.76772	39.0	1.53543	69	2.71654
.9	.03543	20.0	.78740	39.5	1.55512	70	2.75591
1.0	.03937	20.5	.80709	40.0	1.57480	71	2.79528
1.5	.05906	21.0	.82677	40.5	1.59449	72	2.83465
2.0	.07874	21.5	.84646	41.0	1.61417	73	2.87402
2.5	.09843	22.0	.86614	41.5	1.63386	74	2.91339
3.0	.11811	22.5	.88583	42.0	1.65354	75	2.95276
3.5	.13780	23.0	.90551	42.5	1.67323	76	2.99213
4.0	.15748	23.5	.92520	43.0	1.69291	77	3.03150
4.5	.17717	24.0	.94488	43.5	1.71260	78	3.07087
5.0	.19685	24.5	.96457	44.0	1.73228	79	3.11024
5.5	.21654	25.0	.98425	44.5	1.75197	80	3.14961
6.0	.23622	25.5	1.00394	45.0	1.77165	81	3.18898
6.5	.25591	26.0	1.02362	45.5	1.79134	82	3.22835
7.0	.27559	26.5	1.04331	46.0	1.81102	83	3.26772
7.5	.29528	27.0	1.06299	46.5	1.83071	84	3.30709
8.0	.31496	27.5	1.08268	47.0	1.85039	85	3.34646
8.5	.33465	28.0	1.10236	47.5	1.87008	86	3.38583
9.0	.35433	28.5	1.12205	48.0	1.88976	87	3.42520
9.5	.37402	29.0	1.14173	49	1.92913	88	3.46457
10.0	.39370	29.5	1.16142	50	1.96850	89	3.50394
10.5	.41339	30.0	1.18110	51	2.00787	90	3.54331
11.0	.43307	30.5	1.20079	52	2.04724	91	3.58268
11.5	.45276	31.0	1.22047	53	2.08661	92	3.62205
12.0	.47244	31.5	1.24016	54	2.12598	93	3.66142
12.5	.49213	32.0	1.25984	55	2.16535	94	3.70079
13.0	.51181	32.5	1.27953	56	2.20472	95	3.74016
13.5	.53150	33.0	1.29921	57	2.24409	96	3.77953
14.0	.55118	33.5	1.31890	58	2.28347	97	3.81890
14.5	.57087	34.0	1.33858	59	2.32284	98	3.85827
15.0	.59055	34.5	1.35827	60	2.36221	99	3.89764
15.5	.61024	35.0	1.37795	61	2.40158	100	3.93701

NOTE: The above table is approximate: 1/25.4 = 0.039370078+.

DIFFERENT STANDARDS FOR WIRE GAGES IN USE IN THE UNITED STATES

Number of Wire Gage	American or Brown & Sharpe	Birmingham or Stubs' Iron Wire	Washburn & Moen, Worcester, Mass.	W. & M. Steel Music Wire	American S. & W. Co.'s Music Wire Gage	Imperial Wire Gage	Stubs' Steel Wire	U. S. Standard Gage for Sheet and Plate Iron and Steel	Number of Wire Gage
00000000	—	—	—	.0083	—	—	—	—	00000000
0000000	—	—	—	.0087	—	—	—	—	0000000
000000	—	—	—	.0095	.004	.464	—	.46875	000000
00000	—	—	—	.010	.005	.432	—	.4375	00000
0000	.460	.454	.3938	.011	.006	.400	—	.40625	0000
000	.40964	.425	.3625	.012	.007	.372	—	.375	000
00	.3648	.380	.3310	.0133	.008	.348	—	.34375	00
0	.32486	.340	.3065	.0144	.009	.324	—	.3125	0
1	.2893	.300	.2830	.0156	.010	.300	.227	.28125	1
2	.25763	.284	.2625	.0166	.011	.276	.219	.265625	2
3	.22942	.259	.2437	.0178	.012	.252	.212	.250	3
4	.20431	.238	.2253	.0188	.013	.232	.207	.234375	4
5	.18194	.220	.2070	.0202	.014	.212	.204	.21875	5
6	.16202	.203	.1920	.0215	.016	.192	.201	.203125	6
7	.14428	.180	.1770	.023	.018	.176	.199	.1875	7
8	.12849	.165	.1620	.0243	.020	.160	.197	.171875	8
9	.11443	.148	.1483	.0256	.022	.144	.194	.15625	9
10	.10189	.134	.1350	.027	.024	.128	.191	.140625	10
11	.090742	.120	.1205	.0284	.026	.116	.188	.125	11
12	.080808	.109	.1055	.0296	.029	.104	.185	.109375	12
13	.071961	.095	.0915	.0314	.031	.092	.182	.09375	13
14	.064084	.083	.0800	.0326	.033	.080	.180	.078125	14
15	.057068	.072	.0720	.0345	.035	.072	.178	.0703125	15
16	.05082	.065	.0625	.036	.037	.064	.175	.0625	16
17	.045257	.058	.0540	.0377	.039	.056	.172	.05625	17
18	.040303	.049	.0475	.0395	.041	.048	.168	.050	18
19	.03589	.042	.0410	.0414	.043	.040	.164	.04375	19
20	.031961	.035	.0348	.0434	.045	.036	.161	.0375	20
21	.028462	.032	.03175	.046	.047	.032	.157	.034375	21
22	.025347	.028	.0286	.0483	.049	.028	.155	.03125	22
23	.022571	.025	.0258	.051	.051	.024	.153	.028125	23
24	.0201	.022	.0230	.055	.055	.022	.151	.025	24
25	.0179	.020	.0204	.0586	.059	.020	.148	.021875	25
26	.01594	.018	.0181	.0626	.063	.018	.146	.01875	26
27	.014195	.016	.0173	.0658	.067	.0164	.143	.0171875	27
28	.012641	.014	.0162	.072	.071	.0149	.139	.015625	28
29	.011257	.013	.0150	.076	.075	.0136	.134	.0140625	29
30	.010025	.012	.0140	.080	.080	.0124	.127	.0125	30
31	.008928	.010	.0132	—	.085	.0116	.120	.0109375	31
32	.00795	.009	.0128	—	.090	.0108	.115	.01015625	32
33	.00708	.008	.0118	—	.095	.0100	.112	.009375	33
34	.006304	.007	.0104	—	—	.0092	.110	.00859375	34
35	.005614	.005	.0095	—	—	.0084	.108	.0078125	35
36	.005	.004	.0090	—	—	.0076	.106	.00703125	36
37	.004453	—	—	—	—	.0068	.103	.006640625	37
38	.003965	—	—	—	—	.0060	.101	.00625	38
39	.003531	—	—	—	—	.0052	.099	—	39
40	.003144	—	—	—	—	.0048	.097	—	40

DECIMAL EQUIVALENTS OF 8ths, 16ths, 32nds, and 64ths of an INCH

8ths

1/8 = .125
1/4 = .250
3/8 = .375
1/2 = .500
5/8 = .625
3/4 = .750
7/8 = .875

16ths

1/16 = .0625
3/16 = .1875
5/16 = .3125
7/16 = .4375
9/16 = .5625
11/16 = .6875
13/16 = .8125
15/16 = .9375

32nds

1/32 = .03125
3/32 = .09375
5/32 = .15625
7/32 = .21875
9/32 = .28125
11/32 = .34375
13/32 = .40625
15/32 = .46875
17/32 = .53125
19/32 = .59375
21/32 = .65625
23/32 = .71875
25/32 = .78125
27/32 = .84375
29/32 = .90625
31/32 = .96875

64ths

1/64 = .015625
3/64 = .046875
5/64 = .078125
7/64 = .109375
9/64 = .140625
11/64 = .171875
13/64 = .203125
15/64 = .234370
17/64 = .265625
19/64 = .296875
21/64 = .328125
23/64 = .359375
25/64 = .390625
27/64 = .421875
29/64 = .453125
31/64 = .484375
33/64 = .515625
35/64 = .546875
37/64 = .578125
39/64 = .609375
41/64 = .640625
43/64 = .671875
45/64 = .703125
47/64 = .734375
49/64 = .765625
51/64 = .796875
53/64 = .828125
55/64 = .859375
57/64 = .890625
59/64 = .921875
61/64 = .953125
63/64 = .984375

DECIMAL EQUIVALENT OF THE NUMBERS OF TWIST DRILL AND STEEL WIRE GAGE

No.	Size of No. in Decimals	No.	Size of No. in Decimals	No.	Size of No. in Decimals	No.	Size of No. in Decimals	No.	Size of No. in Decimals
1	.2280	17	.1730	33	.1130	49	.0730	65	.0350
2	.2210	18	.1695	34	.1110	50	.0700	66	.0330
3	.2130	19	.1660	35	.1100	51	.0670	67	.0320
4	.2090	20	.1610	36	.1065	52	.0635	68	.0310
5	.2055	21	.1590	37	.1040	53	.0595	69	.0292
6	.2040	22	.1570	38	.1015	54	.0550	70	.0280
7	.2010	23	.1540	39	.0995	55	.0520	71	.0260
8	.1990	24	.1520	40	.0980	56	.0465	72	.0250
9	.1960	25	.1495	41	.0960	57	.0430	73	.0240
10	.1935	26	.1470	42	.0935	58	.0420	74	.0225
11	.1910	27	.1440	43	.0890	59	.0410	75	.0210
12	.1890	28	.1405	44	.0860	60	.0400	76	.0200
13	.1850	29	.1360	45	.0820	61	.0390	77	.0180
14	.1820	30	.1285	46	.0810	62	.0380	78	.0160
15	.1800	31	.1200	47	.0785	63	.0370	79	.0145
16	.1770	32	.1160	48	.0760	64	.0360	80	.0135

MILLIMETERS AND FRACTIONS OF MILLIMETERS

1/100 MM. = .0003937 INCH

mm.	inches	mm.	inches	mm.	inches
1/50 =	.00079	26/50 =	.02047	2 =	.07874
2/50 =	.00157	27/50 =	.02126	3 =	.11811
3/50 =	.00236	28/50 =	.02205	4 =	.15748
4/50 =	.00315	29/50 =	.02283	5 =	.19685
5/50 =	.00394	30/50 =	.02362	6 =	.23622
6/50 =	.00472	31/50 =	.02441	7 =	.27559
7/50 =	.00551	32/50 =	.02520	8 =	.31496
8/50 =	.00630	33/50 =	.02598	9 =	.35433
9/50 =	.00709	34/50 =	.02677	10 =	.39370
10/50 =	.00787	35/50 =	.02756	11 =	.43307
11/50 =	.00866	36/50 =	.02835	12 =	.47244
12/50 =	.00945	37/50 =	.02913	13 =	.51181
13/50 =	.01024	38/50 =	.02992	14 =	.55118
14/50 =	.01102	39/50 =	.03071	15 =	.59055
15/50 =	.01181	40/50 =	.03150	16 =	.62992
16/50 =	.01260	41/50 =	.03228	17 =	.66929
17/50 =	.01339	42/50 =	.03307	18 =	.70866
18/50 =	.01417	43/50 =	.03386	19 =	.74803
19/50 =	.01496	44/50 =	.03465	20 =	.78740
20/50 =	.01575	45/50 =	.03543	21 =	.82677
21/50 =	.01654	46/50 =	.03622	22 =	.86614
22/50 =	.01732	47/50 =	.03701	23 =	.90551
23/50 =	.01811	48/50 =	.03780	24 =	.94488
24/50 =	.01890	49/50 =	.03858	25 =	.98425
25/50 =	.01969	1 =	.03937	26 =	1.02362

10 mm.	=	1 centimeter	=	0.3937 inch
10 cm.	=	1 decimeter	=	3.937 inches
10 dm.	=	1 meter	=	39.37 inches
25.4 mm.	=	1 English inch		

TAPERS AND ANGLES

Taper per Foot	Included < Deg.	Included < Min.	Included < Sec.	With Center Line < Deg.	With Center Line < Min.	With Center Line < Sec.	Taper per Inch	Taper per Inch from Center Line
1/8	0	35	48	0	17	54	.010416	.005203
3/16	0	53	44	0	26	52	.015625	.007812
1/4	1	11	36	0	35	48	.020833	.010416
5/16	1	29	30	0	44	45	.026042	.013021
3/8	1	47	24	0	53	42	.031250	.015625
7/16	2	5	18	1	2	39	.036458	.018229
1/2	2	23	10	1	11	35	.041667	.020833
9/16	2	41	4	1	20	32	.046875	.023438
5/8	2	59	42	1	29	51	.052084	.026042
11/16	3	16	54	1	38	27	.057292	.028646
3/4	3	34	44	1	47	22	.062500	.031250
13/16	3	52	38	1	56	19	.067708	.033854
7/8	4	10	32	2	5	16	.072917	.036456
15/16	4	28	24	2	14	12	.078125	.039063
1	4	46	18	2	23	9	.083330	.041667
1 1/4	5	57	48	2	58	54	.104666	.052084
1 1/2	7	9	10	3	34	35	.125000	.062500
1 3/4	8	20	26	4	10	13	.145833	.072917
2	9	31	36	4	45	48	.166666	.083332
2 1/2	11	53	36	5	56	48	.208333	.104166
3	14	15	0	7	7	30	.250000	.125000
3 1/2	16	35	40	8	17	50	.291666	.145833
4	18	55	28	9	27	44	.333333	.166666
4 1/2	21	14	2	10	37	1	.375000	.187500
5	23	32	12	11	46	6	.416666	.208333
6	28	4	2	14	2	1	.500000	.250000

AMERICAN NATIONAL COARSE AND FINE THREAD DIMENSIONS AND TAP DRILL SIZES

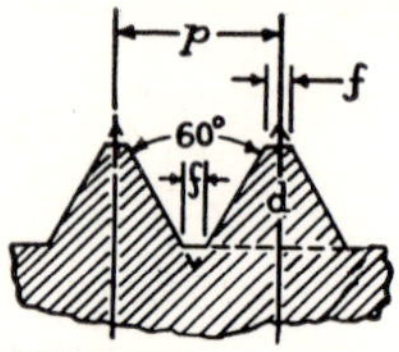

$$p = \text{pitch} = \frac{1}{\text{No. of threads per inch}}$$

$$d = \text{depth} = p \times .649519$$

$$f = \text{flat} = \frac{p}{8}$$

Size	Threads per Inch NC	Threads per Inch NF	Threads per Inch NS	Outside Diameter Inches	Pitch Diameter Inches	Root Diameter Inches	Tap Drill Approx. 75% Full Thread	Decimal Equivalent of Tap Drill
0	—	80	—	.0600	.0519	.0438	3/64	.0469
1	—	—	56	.0730	.0614	.0498	54	.0550
1	64	—	—	.0730	.0629	.0527	53	.0595
1	—	72	—	.0730	.0640	.0550	53	.0595
2	56	—	—	.0860	.0744	.0628	50	.0700
2	—	64	—	.0860	.0759	.0657	49	.0730
3	48	—	—	.0990	.0855	.0719	47	.0785
3	—	56	—	.0990	.0874	.0758	45	.0820
4	—	—	32	.1120	.0917	.0714	45	.0820
4	—	—	36	.1120	.0940	.0759	44	.0860
4	40	—	—	.1120	.0958	.0795	43	.0890
4	—	48	—	.1120	.0985	.0849	42	.0935
5	—	—	36	.1250	.1070	.0889	40	.0980
5	40	—	—	.1250	.1088	.0925	38	.1015
5	—	44	—	.1250	.1102	.0955	37	.1040
6	32	—	—	.1380	.1177	.0974	36	.1065
6	—	—	36	.1380	.1200	.1019	34	.1110
6	—	40	—	.1380	.1218	.1055	33	.1130
8	—	—	30	.1640	.1423	.1207	30	.1285
8	32	—	—	.1640	.1437	.1234	29	.1360
8	—	36	—	.1640	.1460	.1279	29	.1360
8	—	—	40	.1640	.1478	.1315	28	.1405
10	24	—	—	.1900	.1629	.1359	25	.1495
10	—	—	28	.1900	.1668	.1436	23	.1540
10	—	—	30	.1900	.1684	.1467	22	.1570
10	—	32	—	.1900	.1697	.1494	21	.1590
12	24	—	—	.2160	.1889	.1619	16	.1770
12	—	28	—	.2160	.1928	.1696	14	.1820
12	—	—	32	.2160	.1957	.1754	13	.1850
1/4	20	—	—	.2500	.2175	.1850	7	.2010
1/4	—	28	—	.2500	.2268	.2036	3	.2130
5/16	18	—	—	.3125	.2764	.2403	F	.2570
5/16	—	24	—	.3125	.2854	.2584	I	.2720
3/8	16	—	—	.3750	.3344	.2938	5/16	.3125
3/8	—	24	—	.3750	.3479	.3209	Q	.3320
7/16	14	—	—	.4375	.3911	.3447	U	.3680
7/16	—	20	—	.4375	.4050	.3726	25/64	.3906
1/2	13	—	—	.5000	.4500	.4001	27/64	.4219
1/2	—	20	—	.5000	.4675	.4351	29/64	.4531
9/16	12	—	—	.5625	.5084	.4542	31/64	.4844
9/16	—	18	—	.5625	.5264	.4903	33/64	.5156
5/8	11	—	—	.6250	.5660	.5069	17/32	.5312
5/8	—	18	—	.6250	.5889	.5528	37/64	.5781
3/4	10	—	—	.7500	.6850	.6201	21/32	.6562
3/4	—	16	—	.7500	.7094	.6688	11/16	.6875
7/8	9	—	—	.8750	.8028	.7307	49/64	.7656

AMERICAN NATIONAL COARSE AND FINE THREAD DIMENSIONS AND TAP DRILL SIZES

Size	Threads per Inch NC	Threads per Inch NF	NS	Outside Diameter Inches	Pitch Diameter Inches	Root Diameter Inches	Tap Drill Approx. 75% Full Thread	Decimal Equivalent of Tap Drill
7/8	—	14	—	.8750	.8286	.7822	13/16	.8125
7/8	—	—	18	.8750	.8389	.8028	53/64	.8281
1	8	—	—	1.0000	.9188	.8376	7/8	.8750
1	—	14	—	1.0000	.9536	.9072	15/16	.9375
1 1/8	7	—	—	1.1250	1.0322	.9394	63/64	.9844
1 1/8	—	12	—	1.1250	1.0709	1.0168	1 3/64	1.0469
1 1/4	7	—	—	1.2500	1.1572	1.0644	1 7/64	1.1094
1 1/4	—	12	—	1.2500	1.1959	1.1418	1 11/64	1.1719
1 3/8	6	—	—	1.3750	1.2667	1.1585	1 7/32	1.2187
1 3/8	—	12	—	1.3750	1.3209	1.2668	1 19/64	1.2969
1 1/2	6	—	—	1.5000	1.3917	1.2835	1 11/32	1.3437
1 1/2	—	12	—	1.5000	1.4459	1.3918	1 27/64	1.4219
1 3/4	5	—	—	1.7500	1.6201	1.4902	1 9/16	1.5625
2	4 1/2	—	—	2.0000	1.8557	1.7113	1 25/32	1.7812
2 1/4	4 1/2	—	—	2.2500	2.1057	1.9613	2 1/32	2.0313
2 1/2	4	—	—	2.5000	2.3376	2.1752	2 1/4	2.2500
2 3/4	4	—	—	2.7500	2.5876	2.4252	2 1/2	2.5000
3	4	—	—	3.0000	2.8376	2.6752	2 3/4	2.7500
3 1/4	4	—	—	3.2500	3.0876	2.9252	3	3.0000
3 1/2	4	—	—	3.5000	3.3376	3.1752	3 1/4	3.2500
3 3/4	4	—	—	3.7500	3.5876	3.4252	3 1/2	3.5000
4	4	—	—	4.0000	3.3786	3.6752	3 3/4	3.7500

LETTER SIZES OF DRILLS

Diameter Inches	Decimals of 1 Inch	Diameter Inches	Decimals of 1 Inch
A 15/64	.234	N	.302
B	.238	O 5/16	.316
C	.242	P 21/64	.323
D	.246	Q	.332
E 1/4	.250	R 11/32	.339
F	.257	S	.348
G	.261	T 23/64	.358
H 17/64	.266	U	.368
I	.272	V 3/8	.377
J	.277	W 25/64	.386
K 9/32	.281	X	.397
L	.290	Y 13/32	.404
M 19/64	.295	Z	.413

DOUBLE DEPTH OF THREADS

Threads per Inch N	V Threads D D	Am. Nat. Form D D U. S. Std.	Whitworth Standard D D	Threads per Inch N	V Threads D D	Am. Nat. Form D D U.S. Std.	Whitworth Standard D D
2	.86650	.64950	.64000	28	.06185	.04639	.04571
2¼	.77022	.57733	.56888	30	.05773	.04330	.04266
2⅜	.72960	.54694	.53894	32	.05412	.04059	.04000
2½	.69320	.51960	.51200	34	.05097	.03820	.03764
2⅝	.66015	.49485	.48761	36	.04811	.03608	.03555
2¾	.63019	.47236	.46545	38	.04560	.03418	.03368
2⅞	.60278	.45182	.44521	40	.04330	.03247	.03200
3	.57733	.43300	.42666	42	.04126	.03093	.03047
3¼	.53323	.39966	.39384	44	.03936	.02952	.03136
3½	.49485	.37114	.36571	46	.03767	.02823	.02782
4	.43300	.32475	.32000	48	.03608	.02706	.02666
4½	.38488	.28869	.28444	50	.03464	.02598	.02560
5	.34660	.25980	.25600	52	.03332	.02498	.02461
5½	.31490	.23618	.23272	54	.03209	.02405	.02370
6	.28866	.21650	.21333	56	.03093	.02319	.02285
7	.24742	.18557	.18285	58	.02987	.02239	.02206
8	.21650	.16237	.16000	60	.02887	.02165	.02133
9	.19244	.14433	.14222	62	.02795	.02095	.02064
10	.17320	.12990	.12800	64	.02706	.02029	.02000
11	.15745	.11809	.11636	66	.02625	.01968	.01939
11½	.15069	.11295	.11121	68	.02548	.01910	.01882
12	.14433	.10825	.10666	70	.02475	.01855	.01728
13	.13323	.09992	.09846	72	.02407	.01804	.01782
14	.12357	.09278	.09142	74	.02341	.01752	.01729
15	.11555	.08660	.08533	76	.02280	.01714	.01673
16	.10825	.08118	.08000	78	.02221	.01665	.01641
18	.09622	.07216	.07111	80	.02166	.01623	.01600
20	.08660	.06495	.06400	82	.02113	.01584	.01560
22	.07872	.05904	.05818	84	.02063	.01546	.01523
24	.07216	.05412	.05333	86	.02015	.01510	.01476
26	.06661	.04996	.04923	88	.01957	.01476	.01454
27	.06418	.04811	.04740	90	.01925	.01443	.01422

$$DD = \frac{1.733}{N} \text{ For V Thread}$$

$$DD = \frac{1.299}{N} \text{ For American Nat. Form, U. S. Std.}$$

$$DD = \frac{1.28}{N} \text{ For Whitworth Standard}$$

RULES RELATIVE TO THE CIRCLE, ETC.

To Find Circumference:

Multiply diameter by	3.1416	Or divide diameter by	0.3183

To Find Diameter:

Multiply circumference by	0.3183	Or divide circumference by	3.1416

To Find Radius:

Multiply circumference by	0.15915	Or divide circumference by	6.28318

To Find Side of an Inscribed Square:

Multiply diameter by	0.7071		
Or multiply circumference by	0.2251	Or divide circumference by	4.4428

To Find Side of an Equal Square:

Multiply diameter by	0.8862	Or divide diameter by	1.1284
Or multiply circumference by	0.2821	Or divide circumference by	3.545

Square:

A side multiplied by	1.4142	equals diameter of its circumscribing circle.
A side multiplied by	4.443	equals circumference of its circumscribing circle.
A side multiplied by	1.128	equals diameter of an equal circle.
A side multiplied by	3.547	equals circumference of an equal circle.
Square inches multiplied by	1.273	equals circle inches of an equal circle.

To Find the Area of a Circle:

Multiply circumference by one quarter of the diameter.

Or multiply the square of circumference by	0.07958
Or multiply the square of ½ diameter by	3.14160

To Find the Surface of a Sphere or Globe:

Multiply the diameter by the circumference.

Or multiply the square of diameter by	3.1416
Or multiply four times the square of radius by	3.1416

To Find the Weight of Brass and Copper Sheets, Rods and Bars:

Calculate the volume, i.e., number of cubic inches in the piece and multiply by weight per cubic inch.

Brass, 0.2972
Copper, 0.3212

Or multiply the length by the breadth (in feet) and product by weight in pounds per square foot.

METRIC CONVERSION TABLE

Millimeters	×	.03937	= Inches
Millimeters	=	25.400	× Inches
Meters	×	3.2809	= Feet
Meters	=	.3048	× Feet
Kilometers	×	.621377	= Miles
Kilometers	=	1.6093	× Miles
Square centimeters	×	.15500	= Square inches
Square centimeters	=	6.4515	× Square inches
Square meters	×	10.76410	= Square feet
Square meters	=	.09290	× Square feet
Square kilometers	×	247.1098	= Acres
Square kilometers	=	.00405	× Acres
Hectares	×	2.471	= Acres
Hectares	=	.4047	× Acres
Cubic centimeters	×	.061025	= Cubic inches
Cubic centimeters	=	16.3866	× Cubic inches
Cubic meters	×	35.3156	= Cubic feet
Cubic meters	=	.02832	× Cubic feet
Cubic meters	×	1.308	= Cubic yards
Cubic meters	=	.765	× Cubic yards
Liters	×	61.023	= Cubic inches
Liters	=	.01639	× Cubic inches
Liters	×	.26418	= U. S. gallons
Liters	=	3.7854	× U. S. gallons
Grams	×	15.4324	= Grains
Grams	=	.0648	× Grains
Grams	×	.03527	= Ounces, avoirdupois
Grams	=	28.3495	× Ounces, avoirdupois
Kilograms	×	2.2046	= Pounds
Kilograms	=	.4536	× Pounds
Kilograms per square centimeter	×	14.2231	= Pounds per square inch
Kilograms per square centimeter	=	.0703	× Pounds per square inch
Kilograms per cubic meter	×	.06243	= Pounds per cubic foot
Kilograms per cubic meter	=	16.01890	× Pounds per cubic foot
Metric tons (1,000 kilograms)	×	1.1023	= Tons (2,000 pounds)
Metric tons	=	.9072	× Tons (2,000 pounds)
Kilowatts	×	1.3405	= Horse-powers
Kilowatts	=	.746	× Horse-powers
Calories	×	3.9683	= B. T. units
Calories	=	.2520	× B. T. units

By courtesy of the American Machinist, New York.

TABLES OF DECIMAL EQUIVALENTS

Of 7ths, 14ths, and 28ths of an Inch								Of 6ths, 12ths, and 24ths of an Inch							
7th	14th	28th	Decimal	7th	14th	28th	Decimal	6th	12th	24th	Decimal	6th	12th	24th	Decimal
		1	.035714			15	.535714			1	.041667			13	.541666
	1		.071429	4			.571429		1		.083333		7		.583333
		3	.107143			17	.607143			3	.125			15	.625
1			.142857		9		.642867	1			.166666	4			.666666
		5	.178571			19	.678571			5	.208333			17	.708333
	3		.214286	5			.714286		3		.25		9		.75
		7	.25			21	.75			7	.291666			19	.791666
2			.285714		11		.785714	2			.333333	5			.833333
		9	.321429			23	.821429			9	.375			21	.875
	5		.357143	6			.857143		5		.416666		11		.916666
		11	.392857			25	.892857			11	.458333			23	.958333
3			.428571		13		.928571	3			.5				
		13	.464286			27	.964286								
	7		.5												

TABLES FOR COMPUTING WEIGHT OF CAST STEEL

WEIGHT IN POUNDS OF A LINEAL FOOT OF ROUND, SQUARE, AND OCTAGON STEEL

Size in Inches	Round	Octagon	Square	Size in Inches	Round	Octagon	Square
1/16	.010	.011	.013	2 1/2	16.79	17.71	21.37
1/8	.042	.044	.053	2 5/8	18.51	19.52	23.56
3/16	.094	.099	.120	2 3/4	20.31	21.42	25.86
1/4	.168	.177	.214	2 7/8	22.20	23.41	28.27
5/16	.262	.277	.334	3	24.17	25.50	30.78
3/8	.378	.398	.481	3 1/8	26.23	27.66	33.40
7/16	.514	.542	.655	3 1/4	28.37	29.92	36.12
1/2	.671	.708	.855	3 3/8	30.59	32.27	38.95
9/16	.850	.896	1.082	3 1/2	32.90	34.70	41.89
5/8	1.049	1.107	1.336	3 5/8	35.29	37.23	44.94
11/16	1.270	1.339	1.616	3 3/4	37.77	39.84	48.09
3/4	1.511	1.594	1.924	3 7/8	40.33	42.54	51.35
13/16	1.773	1.870	2.258	4	42.97	45.33	54.72
7/8	2.056	2.169	2.618	4 1/4	48.51	51.17	61.77
15/16	2.361	2.490	3.006	4 1/2	54.39	57.37	69.25
1	2.686	2.833	3.420	4 3/4	60.60	63.92	77.16
1 1/8	3.399	3.585	4.328	5	67.15	70.83	85.50
1 1/4	4.197	4.427	5.344	5 1/4	74.03	78.08	94.26
1 3/8	5.078	5.356	6.466	5 1/2	81.25	85.70	103.45
1 1/2	6.044	6.374	7.695	5 3/4	88.80	93.67	113.07
1 5/8	7.093	7.481	9.031	6	96.69	101.99	123.12
1 3/4	8.226	8.674	10.474	7	131.61	138.82	167.58
1 7/8	9.443	9.960	12.023	8	171.90	181.32	218.88
2	10.744	11.332	13.680	9	217.57	229.48	277.02
2 1/8	12.129	12.793	15.443	10	268.60	283.31	342.00
2 1/4	13.598	14.343	17.314	11	325.01	342.80	413.82
2 3/8	15.151	15.981	19.291	12	386.79	407.97	492.48

MISCELLANEOUS MEASUREMENTS

Measures of Length

1 mile = 1760 yards = 5280 feet.
1 yard = 3 feet = 36 inches.
1 foot = 12 inches.

The following measures of length are also used occasionally:
1 mil = 0.001 inch. 1 fathom = 2 yards = 6 feet.
1 rod = 5.5 yards = 16.5 feet. 1 hand = 4 inches. 1 span = 9 inches.

Surveyor's Measure

1 mile = 8 furlongs = 80 chains.
1 furlong = 10 chains = 220 yards.
1 chain = 4 rods = 22 yards = 66 feet = 100 links.
1 link = 7.92 inches.

Nautical Measure

1 league = 3 nautical miles.
1 nautical mile (knot) = 6080.26 feet = 1.1516 statute mile.
One degree at the equator = 60 nautical miles = 69.168 statute miles.
360 degrees = 21,600 nautical miles = 24,874.5 statute miles = circumference of earth at the equator.

Square Measure

1 square mile = 640 acres = 6400 square chains.
1 acre = 10 square chains = 4840 square yards = 43,560 square feet.
1 square chain = 16 square rods = 484 square yards = 4356 square feet.
1 square rod = 30.25 square yards = 272.25 square feet = 625 square links.
1 square yard = 9 square feet.
1 square foot = 144 square inches.
An acre is equal to a square, the side of which is 208.7 feet.

TAP DRILL SIZES

75% Depth of Thread

A bolt inserted in an ordinary nut, which has only one half of a full depth of thread, will break before stripping the thread. Also a full depth of thread, while very difficult to obtain, is only about 5% stronger than a 75% depth.

These tables give the exact size of the hole, expressed in decimals, that will produce a 75% depth of thread, and also the nearest regular stock drill to this size. Holes produced by these drills are considered close enough for any commercial tapping.

$$\text{Diameter of Tap, Minus } \frac{.974}{\text{No. threads per Inch}} = \text{Diameter of Holes}$$

TAP DRILL SIZES—75% Depth Thread
Machine Screw Threads

Tap Size	Threads per Inch	Diameter Hole	Drill	Tap Size	Threads per Inch	Diameter Hole	Drill
0	★80	.048	3/64	10	32	.160	21
1	★72	.060	53	10	★30	.158	22
1	64	.058	53	10	24	.149	25
2	★64	.071	50	12	★28	.181	14
2	56	.069	50	12	24	.175	16
3	★56	.082	45	14	★24	.201	7
3	48	.079	47	14	20	.193	10
4	★48	.092	42	16	★22	.224	2
4	40	.088	43	16	20	.219	7/32
4	36	.085	44	16	18	.214	3
5	★44	.103	37	18	★20	.245	D
5	40	.101	38	18	18	.240	B
5	36	.098	40	20	★20	.271	I
6	★40	.114	33	20	18	.266	17/64
6	36	.111	34	22	★18	.292	L
6	32	.108	36	22	16	.285	9/32
7	★36	.124	1/8	24	18	.318	O
7	32	.121	31	24	★16	.311	5/16
7	30	.119	31	26	★16	.337	R
8	★36	.137	29	26	14	.328	21/64
8	32	.134	29	28	16	.363	23/64
8	30	.132	30	28	★14	.354	T
9	★32	.147	26	30	16	.389	25/64
9	30	.145	27	30	★14	.380	V
9	24	.136	29				

★A. S. M. E. Standard

Tap Drill Sizes

75% Depth Thread

American National Form **U. S. and S. A. E. Standard**

Tap Size	Threads per Inch	Diam. Hole	Drill	Tap Size	Threads per Inch	Diam. Hole	Drill	Tap Size	Threads per Inch	Diam. Hole	Drill
1/16	72	.049	3/64	1/4	32	.220	7/32	7/8	*14	.805	13/16
** 1/16	64	.047	3/64	1/4	*28	.215	3	7/8	12	.794	51/64
1/16	60	.046	56	1/4	27	.214	3	** 7/8	9	.767	49/64
5/64	72	.065	52	1/4	24	.209	4	15/16	12	.856	55/64
5/64	64	.063	1/16	** 1/4	20	.201	7	** 15/16	9	.829	53/64
** 5/64	60	.062	1/16	5/16	32	.282	9/32	1	27	.964	31/32
5/64	56	.061	53	5/16	27	.276	J	1	*14	.930	15/16
3/32	60	.077	5/64	5/16	*24	.272	I	1	12	.919	59/64
3/32	56	.076	48	5/16	20	.264	17/64	**1	8	.878	7/8
** 3/32	50	.074	49	** 5/16	18	.258	F	1 1/16	8	.941	15/16
3/32	48	.073	49	3/8	27	.339	R	1 1/8	*12	1.044	1 3/64
7/64	56	.092	42	3/8	*24	.334	Q	**1 1/8	7	.986	63/64
7/64	50	.090	43	3/8	20	.326	21/64	1 3/16	7	1.048	1 3/64
** 7/64	48	.089	43	** 3/8	16	.314	5/16	1 1/4	*12	1.169	1 11/64
1/8	48	.105	36	7/16	27	.401	Y	**1 1/4	7	1.111	1 7/64
** 1/8	40	.101	38	7/16	24	.397	X	1 5/16	7	1.173	1 11/64
1/8	36	.098	40	7/16	*20	.389	25/64	1 3/8	*12	1.294	1 19/64
1/8	32	.095	3/32	** 7/16	14	.368	U	**1 3/8	6	1.213	1 7/32
** 9/64	40	.116	32	1/2	27	.464	15/32	1 1/2	*12	1.419	1 27/64
9/64	36	.114	33	1/2	24	.460	29/64	**1 1/2	6	1.338	1 11/32
9/64	32	.110	35	1/2	*20	.451	29/64	**1 5/8	5 1/2	1.448	1 29/64
5/32	40	.132	30	** 1/2	13	.425	27/64	**1 3/4	5	1.555	1 9/16
** 5/32	36	.129	30	1/2	12	.419	27/64	**1 7/8	5	1.680	1 11/16
5/32	32	.126	1/8	9/16	27	.526	17/32	**2	4 1/2	1.783	1 25/32
11/64	36	.145	27	9/16	*18	.508	33/64	**2 1/8	4 1/2	1.909	1 29/32
**11/64	32	.141	9/64	** 9/16	12	.481	31/64	**2 1/4	4 1/2	2.034	2 1/32
3/16	36	.161	20	5/8	27	.589	19/32	**2 3/8	4	2.131	2 1/8
3/16	32	.157	22	5/8	*18	.571	37/64	**2 1/2	4	2.256	2 1/4
3/16	30	.155	23	5/8	12	.544	35/64	**2 5/8	4	2.381	2 3/8
** 3/16	24	.147	26	** 5/8	11	.536	17/32	**2 3/4	4	2.506	2 1/2
13/64	32	.173	17	11/16	*16	.627	5/8	**2 7/8	3 1/2	2.597	2 19/32
13/64	30	.171	11/64	**11/16	11	.599	19/32	**3	3 1/2	2.722	2 23/32
**13/64	24	.163	20	3/4	27	.714	23/32	**3 1/8	3 1/2	2.847	2 27/32
7/32	32	.188	12	3/4	*16	.689	11/16	**3 1/4	3 1/2	2.972	2 31/32
7/32	28	.184	13	3/4	12	.669	43/64	**3 3/8	3 1/4	3.075	3 1/16
** 7/32	24	.178	16	** 3/4	10	.653	21/32	**3 1/2	3 1/4	3.200	3 3/16
15/64	32	.204	6	13/16	12	.731	47/64	**3 5/8	3 1/4	3.325	3 5/16
15/64	28	.200	8	**13/16	10	.715	23/32	**3 3/4	3	3.425	3 7/16
**15/64	24	.194	10	7/8	27	.839	27/32	**3 7/8	3	3.550	3 9/16
				7/8	*18	.821	53/64	**4	3	3.675	3 11/16

*S. A. E. Standard **U. S. Standard

THE METRIC SYSTEM OF MEASUREMENT

Measures of Length

1 Millimeter (mm.) = ...0.03937079 inch, or about 1/25 inch

10 Millimeters = 1 Centimeter (cm.) = ...0.3937079 inch

10 Centimeters = 1 Decimeter (dm.) = ...3.937079 inch

10 Decimeters = 1 Meter (m.) = 39.37079 inches, 3.2808992 feet, or 1.09361 yards

10 Meters = 1 Decameter (Dm.) = ...32.808992 feet

10 Decameters = 1 Hectometer (Hm.) = ...19.927817 rods

10 Hectometers = 1 Kilometer (Km.) = ...1093.61 yards, or 0.6213824 mile

10 Kilometers = 1 Myriameter (Mm.) = ...6.213824 miles

1 inch = 2.54 cm., 1 foot = 0.3048 m., 1 yard = 0.9144 m., 1 rod = 0.5029 Dm., 1 mile = 1.6093 Km.

Measures of Weight

1 Gram (g.) = 15.4324874 gr. Troy, or 0.03215 oz. Troy, or 0.03527398 oz. avoirdupois

10 Grams = 1 Decagram (Dg.) = ...0.3527397 oz. avoirdupois

10 Decagrams = 1 Hectogram (Hg.) = ...3.527398 oz. avoirdupois

10 Hectograms = 1 Kilogram (Kg.) = ...2.20462125 lbs.

1000 Kilograms = 1 Ton (T.) = 2204.62125 lb., or 1.1023 tons of 2000 lb., or 0.9842 ton of 2240 lb., or 19.68 cwt.

1 grain = 0.0648 g., 1 oz. avoirdupois = 28.35 g., 1 lb. = 0.4536 Kg., 1 ton (2000 lb.) = 0.9072 T., 1 ton (2240 lb.) = 1.016 T., or 1016 Kg.

Measures of Capacity

1 Liter (l.) = 1 cubic decimeter = 61.0270515 cubic in., or 0.03531 cu. ft., or 1.0567 liquid qt., or 0.908 dry qt., or 0.26417 Amer. gal.

10 Liters = 1 Decaliter (Dl.) = 2.6417 gal., or 1.135 pk.

10 Decaliters = 1 Hectoliter (Hl.) = 2.8375 bu.

10 Hectoliters = 1 Kiloliter (Kl.) = 61027.0515 cu. in., or 28.375 bu.

1 cu. foot = 28.317 l., 1 gallon (American) = 3.785 l., 1 gallon (British) = 4.543 l.

TRIGONOMETRY TABLES

Deg.	Sine	Cosine	Tan.	Cotan.	Secant	Cosec.
1	.01745	.99985	.01745	57.290	1.0001	57.299
2	.03490	.99939	.03492	28.636	1.0006	28.654
3	.05234	.99863	.05241	19.081	1.0014	19.107
4	.06976	.99756	.06993	14.301	1.0024	14.335
5	.08715	.99619	.08749	11.430	1.0038	11.474
6	.10453	.99452	.10510	9.5144	1.0055	9.5668
7	.12187	.99255	.12278	8.1443	1.0075	8.2055
8	.13917	.99027	.14054	7.1154	1.0098	7.1853
9	.15643	.98769	.15838	6.3137	1.0125	6.3924
10	.17365	.98481	.17633	5.6713	1.0154	5.7588
11	.19081	.98163	.19438	5.1445	1.0187	5.2408
12	.20791	.97815	.21256	4.7046	1.0223	4.8097
13	.22495	.97437	.23087	4.3315	1.0263	4.4454
14	.24192	.97029	.24933	4.0108	1.0306	4.1336
15	.25882	.96592	.26795	3.7320	1.0353	3.8637
16	.27564	.96126	.28674	3.4874	1.0403	3.6279
17	.29237	.95630	.30573	3.2708	1.0457	3.4203
18	.30902	.95106	.32492	3.0777	1.0515	3.2361
19	.32557	.94552	.34433	2.9042	1.0576	3.0715
20	.34202	.93969	.36397	2.7475	1.0642	2.9238
21	.35837	.93358	.38386	2.6051	1.0711	2.7904
22	.37461	.92718	.40403	2.4751	1.0785	2.6695
23	.39073	.92050	.42447	2.3558	1.0864	2.5593
24	.40674	.91354	.44523	2.2460	1.0946	2.4586
25	.42262	.90631	.46631	2.1445	1.1034	2.3662
26	.43837	.89879	.48773	2.0503	1.1126	2.2812
27	.45399	.89101	.50952	1.9626	1.1223	2.2027
28	.46947	.88295	.53171	1.8807	1.1326	2.1300
29	.48481	.87462	.55431	1.8040	1.1433	2.0627
30	.50000	.86603	.57735	1.7320	1.1547	2.0000
31	.51504	.85717	.60086	1.6643	1.1666	1.9416
32	.52992	.84805	.62487	1.6003	1.1792	1.8871
33	.54464	.83867	.64941	1.5399	1.1924	1.8361
34	.59919	.82904	.67451	1.4826	1.2062	1.7883
35	.57358	.81915	.70021	1.4281	1.2208	1.7434
36	.58778	.80902	.72654	1.3764	1.2361	1.7013
37	.60181	.79863	.75355	1.3270	1.2521	1.6616
38	.61566	.78801	.78128	1.2799	1.2690	1.6243
39	.62932	.77715	.80978	1.2349	1.2867	1.5890
40	.64279	.76604	.83910	1.1917	1.3054	1.5557
41	.65606	.75471	.86929	1.1504	1.3250	1.5242
42	.66913	.74314	.90040	1.1106	1.3456	1.4945
43	.68200	.73135	.93251	1.0724	1.3673	1.4663
44	.69466	.71934	.96569	1.0355	1.3902	1.4395
45	.70711	.70711	1.0000	1.0000	1.4142	1.4142

TRIGONOMETRY TABLES

Deg.	Sine	Cosine	Tan.	Cotan.	Secant	Cosec.
46	.71934	.69466	1.0355	.96569	1.4395	1.3902
47	.73135	.68200	1.0724	.93251	1.4663	1.3673
48	.74314	.66913	1.1106	.90040	1.4945	1.3456
49	.75471	.65606	1.1504	.86929	1.5242	1.3250
50	.76604	.64279	1.1917	.83910	1.5557	1.3054
51	.77715	.62932	1.2342	.80978	1.5890	1.2867
52	.78801	.61566	1.2799	.78128	1.6243	1.2690
53	.79863	.60181	1.3270	.75355	1.6616	1.2521
54	.80902	.58778	1.3764	.72654	1.7013	1.2361
55	.81915	.57358	1.4281	.70021	1.7434	1.2208
56	.82904	.55919	1.4826	.67451	1.7883	1.2062
57	.83867	.54464	1.5399	.64941	1.8361	1.1924
58	.84805	.52992	1.6003	.62487	1.8871	1.1792
59	.85717	.51504	1.6643	.60086	1.9416	1.1666
60	.86603	.50000	1.7320	.57735	2.0000	1.1547
61	.87462	.48481	1.8040	.55431	2.0627	1.1433
62	.88295	.46947	1.8807	.53171	2.1300	1.1326
63	.89101	.45399	1.9626	.50952	2.2027	1.1223
64	.89879	.43837	2.0503	.48773	2.2812	1.1126
65	.90631	.42262	2.1445	.46631	2.3662	1.1034
66	.91354	.40674	2.2460	.44523	2.4586	1.0946
67	.92050	.39073	2.3558	.42447	2.5593	1.0864
68	.92718	.37461	2.4751	.40403	2.6695	1.0785
69	.93358	.35837	2.6051	.38386	2.7904	1.0711
70	.93969	.34202	2.7475	.36397	2.9238	1.0642
71	.94552	.32557	2.9042	.34433	3.0715	1.0576
72	.95106	.30902	3.0777	.32492	3.2361	1.0515
73	.95630	.29237	3.2708	.30573	3.4203	1.0457
74	.96126	.27564	3.4874	.28674	3.6279	1.0403
75	.96592	.25882	3.7320	.26795	3.8637	1.0353
76	.97029	.24192	4.0108	.24833	4.1336	1.0306
77	.97437	.22495	4.3315	.23087	4.4454	1.0263
78	.97815	.20791	4.7046	.21256	4.8097	1.0223
79	.98163	.19081	5.1445	.19438	5.2408	1.0187
80	.98481	.17365	5.6713	.17633	5.7588	1.0154
81	.98769	.15643	6.3137	.15838	6.3924	1.0125
82	.99027	.13917	7.1154	.14054	7.1853	1.0098
83	.99255	.12187	8.1443	.12278	8.2055	1.0075
84	.99452	.10453	9.5144	.10510	9.5668	1.0055
85	.99619	.08715	11.430	.08749	11.474	1.0038
86	.99756	.06976	14.301	.06993	14.335	1.0024
87	.99863	.05234	19.081	.05241	19.107	1.0014
88	.99939	.03490	28.636	.03492	28.654	1.0006
89	.99985	.01745	57.290	.01745	57.299	1.0001

TABLES OF CONSTANTS FOR 5-INCH SINE BAR

WHEN USING 10-INCH SINE BAR MULTIPLY CONSTANT BY TWO

M	0°	1°	2°	3°	4°	5°
0	0.00000	0.08725	0.17450	0.26170	0.34880	0.43580
1	.00145	.08870	.17595	.26315	.35025	.43725
2	.00290	.09015	.17740	.26460	.35170	.43870
3	.00435	.09160	.17885	.26605	.35315	.44015
4	.00580	.09310	.18030	.26750	.35460	.44155
5	0.00725	0.09455	0.18175	0.26895	0.35605	0.44300
6	.00875	.09600	.18320	.27040	.35750	.44445
7	.01020	.09745	.18465	.27185	.35895	.44590
8	.01165	.09890	.18615	.27330	.36040	.44735
9	.01310	.10035	.18760	.27475	.36185	.44880
10	0.01455	0.10180	0.18905	0.27620	0.36330	0.45025
11	.01600	.10325	.19050	.27765	.36475	.45170
12	.01745	.10470	.19195	.27910	.36620	.45315
13	.01890	.10615	.19340	.28055	.36765	.45460
14	.02035	.10760	.19485	.28200	.36910	.45605
15	0.02180	0.10905	0.19630	0.28345	0.37055	0.45750
16	.02325	.11055	.19775	.28490	.37200	.45895
17	.02475	.11200	.19920	.28635	.37345	.46040
18	.02620	.11345	.20065	.28780	.37490	.46185
19	.02765	.11490	.20210	.28925	.37635	.46330
20	0.02910	0.11635	0.20355	0.29070	0.37780	0.46475
21	.03055	.11780	.20500	.29220	.37925	.46620
22	.03200	.11925	.20645	.29365	.38070	.46765
23	.03345	.12070	.20795	.29510	.38215	.46910
24	.03490	.12215	.20940	.29655	.38360	.47055
25	0.03635	0.12360	0.21085	0.29800	0.38505	0.47200
26	.03780	.12505	.21230	.29945	.38650	.47345
27	.03925	.12650	.21375	.30090	.38795	.47490
28	.04070	.12800	.21520	.30235	.38940	.47635
29	.04220	.12945	.21665	.30380	.39085	.47780
30	0.04365	0.13090	0.21810	0.30525	0.39230	0.47925
31	.04510	.13235	.21955	.30670	.39375	.48070
32	.04655	.13380	.22100	.30815	.39520	.48210
33	.04800	.13525	.22245	.30960	.39665	.48355
34	.04945	.13670	.22390	.31105	.39810	.48500
35	0.05090	0.13815	0.22535	0.31250	0.39955	0.48645
36	.05235	.13960	.22680	.31395	.40100	.48790
37	.05380	.14105	.22825	.31540	.40245	.48935
38	.05525	.14250	.22970	.31685	.40390	.49080
39	.05670	.14395	.23115	.31830	.40535	.49225
40	0.05820	0.14540	0.23265	0.31975	0.40680	0.49370
41	.05965	.14690	.23410	.32120	.40825	.49515
42	.06110	.14835	.23555	.32265	.40970	.49660
43	.06255	.14980	.23700	.32410	.41115	.49805
44	.06400	.15125	.23845	.32555	.41260	.49950
45	0.06545	0.15270	0.23990	0.32700	0.41405	0.50095
46	.06690	.15415	.24135	.32845	.41550	.50240
47	.06835	.15560	.24280	.32990	.41695	.50385
48	.06980	.15705	.24425	.33135	.41840	.50530
49	.07125	.15850	.24570	.33280	.41985	.50675
50	0.07270	0.15995	0.24715	0.33425	0.42130	0.50820
51	.07415	.16140	.24860	.33570	.42275	.50960
52	.07565	.16285	.25005	.33715	.42420	.51105
53	.07710	.16430	.25150	.33865	.42565	.51250
54	.07855	.16580	.25295	.34010	.42710	.51395
55	0.08000	0.16725	0.25440	0.34155	0.42855	0.51540
56	.08145	.16870	.25585	.34300	.43000	.51685
57	.08290	.17015	.25730	.34445	.43145	.51830
58	.08435	.17160	.25875	.34590	.43290	.51975
59	.08580	.17305	.26028	.34735	.43435	.52120
60	0.08725	0.17450	0.26170	0.34880	0.43580	0.52265

TABLES OF CONSTANTS FOR 5-INCH SINE BAR

M	6°	7°	8°	9°	10°	11°
0	0.52265	0.60935	0.69585	0.78215	0.86825	0.95405
1	.52410	.61080	.69730	.78360	.86965	.95545
2	.52555	.61225	.69875	.78505	.87110	.95690
3	.52700	.61370	.70020	.78650	.87255	.95835
4	.52845	.61510	.70165	.78790	.87395	.95975
5	0.52985	0.61655	0.70305	0.78935	0.87540	0.96120
6	.53130	.61800	.70450	.79080	.87685	.96260
7	.53275	.61945	.70595	.79225	.87825	.96405
8	.53420	.62090	.70740	.79365	.87970	.96545
9	.53565	.62235	.70885	.79510	.88115	.96690
10	0.53710	0.62380	0.71025	0.79655	0.88255	0.96830
11	.53855	.62520	.71170	.79795	.88400	.96975
12	.54000	.62665	.71315	.79940	.88540	.97115
13	.54145	.62810	.71460	.80085	.88685	.97260
14	.54290	.62955	.71600	.80230	.88830	.97405
15	0.54435	0.63100	0.71745	0.80370	0.88970	0.97545
16	.54580	.63245	.71890	.80515	.89115	.97690
17	.54725	.63390	.72035	.80660	.89260	.97830
18	.54865	.63530	.72180	.80800	.89400	.97975
19	.55010	.63675	.72320	.80945	.89545	.98115
20	0.55155	0.63820	0.72465	0.81090	0.89685	0.98260
21	.55300	.63965	.72610	.81230	.89830	.98400
22	.55445	.64110	.72755	.81375	.89975	.98545
23	.55590	.64255	.72900	.81520	.90115	.98685
24	.55735	.64400	.73040	.81665	.90260	.98830
25	0.55880	0.64540	0.73185	0.81805	0.90405	0.98970
26	.56025	.64685	.73330	.81950	.90545	.99115
27	.56170	.64830	.73475	.82095	.90690	.99255
28	.56315	.64975	.73615	.82235	.90830	.99400
29	.56455	.65120	.73760	.82380	.90975	.99540
30	0.56600	0.65265	0.73905	0.82525	0.91120	0.99685
31	.56745	.65405	.74050	.82665	.91260	.99825
32	.56890	.65550	.74190	.82810	.91405	.99970
33	.57035	.65695	.74335	.82955	.91545	1.0011
34	.57180	.65840	.74480	.83100	.91690	.0016
35	0.57325	0.65985	0.74625	0.83240	0.91835	1.0039
36	.57470	.66130	.74770	.83385	.91975	.0054
37	.57615	.66270	.74910	.83530	.92120	.0068
38	.57760	.66415	.75055	.83670	.92260	.0082
39	.57900	.66560	.75200	.83815	.92405	.0096
40	0.58045	0.66705	0.75345	0.83960	0.92545	1.0110
41	.58190	.66850	.75485	.84100	.92690	.0125
42	.58335	.66995	.75630	.84245	.92835	.0139
43	.58480	.67135	.75775	.84390	.92975	.0153
44	.58625	.67280	.75920	.84530	.93120	.0168
45	0.58770	0.67425	0.76060	0.84675	0.93260	1.0182
46	.58915	.67570	.76205	.84820	.93405	.0196
47	.59060	.67715	.76350	.84960	.93550	.0210
48	.59200	.67860	.76495	.85105	.93690	.0225
49	.59345	.68000	.76635	.85250	.93835	.0239
50	0.59490	0.68145	0.76780	0.85390	0.93975	1.0253
51	.59635	.68290	.76925	.85535	.94120	.0267
52	.59780	.68435	.77070	.85680	.94260	.0281
53	.59925	.68580	.77210	.85820	.94405	.0296
54	.60070	.68720	.77355	.85965	.94550	.0310
55	0.60215	0.68865	0.77500	0.86110	0.94690	1.0324
56	.60355	.69010	.77645	.86250	.94835	.0338
57	.60500	.69155	.77785	.86395	.94975	.0353
58	.60645	.69300	.77930	.86540	.95120	.0367
59	.60790	.69445	.78075	.86680	.95260	.0381
60	0.60935	0.69585	0.78215	0.86825	0.95405	1.0395

TABLES OF CONSTANTS FOR 5-INCH SINE BAR

M	12°	13°	14°	15°	16°	17°
0	1.0395	1.1247	1.2096	1.2941	1.3782	1.4618
1	.0410	.1261	.2110	.2955	.3796	.4632
2	.0424	.1276	.2124	.2969	.3810	.4646
3	.0438	.1290	.2138	.2983	.3824	.4660
4	.0452	.1304	.2152	.2997	.3838	.4674
5	1.0466	1.1318	1.2166	1.3011	1.3852	1.4688
6	.0481	.1332	.2181	.3025	.3865	.4702
7	.0495	.1346	.2195	.3039	.3879	.4716
8	.0509	.1361	.2209	.3053	.3893	.4730
9	.0523	.1375	.2223	.3067	.3907	.4743
10	1.0538	1.1389	1.2237	1.3081	1.3921	1.4757
11	.0552	.1403	.2251	.3095	.3935	.4771
12	.0566	.1417	.2265	.3109	.3949	.4785
13	.0580	.1431	.2279	.3123	.3963	.4799
14	.0594	.1446	.2293	.3137	.3977	.4813
15	1.0609	1.1460	1.2307	1.3151	1.3991	1.4827
16	.0623	.1474	.2322	.3165	.4005	.4841
17	.0637	.1488	.2336	.3179	.4019	.4855
18	.0651	.1502	.2350	.3193	.4033	.4868
19	.0665	.1516	.2364	.3207	.4047	.4882
20	1.0680	1.1531	1.2378	1.3221	1.4061	1.4896
21	.0694	.1545	.2392	.3235	.4075	.4910
22	.0708	.1559	.2406	.3250	.4089	.4924
23	.0722	.1573	.2420	.3264	.4103	.4938
24	.0737	.1587	.2434	.3278	.4117	.4952
25	1.0751	1.1601	1.2448	1.3292	1.4131	1.4966
26	.0765	.1615	.2462	.3306	.4145	.4980
27	.0779	.1630	.2477	.3320	.4159	.4993
28	.0793	.1644	.2491	.3334	.4173	.5007
29	.0808	.1658	.2505	.3348	.4187	.5021
30	1.0822	1.1672	1.2519	1.3362	1.4201	1.5035
31	.0836	.1686	.2533	.3376	.4214	.5049
32	.0850	.1700	.2547	.3390	.4228	.5063
33	.0864	.1714	.2561	.3404	.4242	.5077
34	.0879	.1729	.2575	.3418	.4256	.5091
35	1.0893	1.1743	1.2589	1.3432	1.4270	1.5104
36	.0907	.1757	.2603	.3446	.4284	.5118
37	.0921	.1771	.2617	.3460	.4298	.5132
38	.0935	.1785	.2631	.3474	.4312	.5146
39	.0949	.1799	.2645	.3488	.4326	.5160
40	1.0964	1.1813	1.2660	1.3502	1.4340	1.5174
41	.0978	.1828	.2674	.3516	.4354	.5188
42	.0992	.1842	.2688	.3530	.4368	.5201
43	.1006	.1856	.2702	.3544	.4382	.5215
44	.1020	.1870	.2716	.3558	.4396	.5229
45	1.1035	1.1884	1.2730	1.3572	1.4410	1.5243
46	.1049	.1898	.2744	.3586	.4423	.5257
47	.1063	.1912	.2758	.3600	.4437	.5271
48	.1077	.1926	.2772	.3614	.4451	.5285
49	.1091	.1941	.2786	.3628	.4465	.5298
50	1.1106	1.1955	1.2800	1.3642	1.4479	1.5312
51	.1120	.1969	.2814	.3656	.4493	.5326
52	.1134	.1983	.2828	.3670	.4507	.5340
53	.1148	.1997	.2842	.3684	.4521	.5354
54	.1162	.2011	.2856	.3698	.4535	.5368
55	1.1176	1.2025	1.2870	1.3712	1.4549	1.5381
56	.1191	.2039	.2884	.3726	.4563	.5395
57	.1205	.2054	.2899	.3740	.4577	.5409
58	.1219	.2068	.2913	.3754	.4591	.5423
59	.1233	.2082	.2927	.3768	.4604	.5437
60	1.1247	1.2096	1.2941	1.3782	1.4618	1.5451

TABLES OF CONSTANTS FOR 5-INCH SINE BAR

M	18°	19°	20°	21°	22°	23°
0	1.5451	1.6278	1.7101	1.7918	1.8730	1.9536
1	.5464	.6292	.7114	.7932	.8744	.9550
2	.5478	.6306	.7128	.7945	.8757	.9563
3	.5492	.6319	.7142	.7959	.8771	.9576
4	.5506	.6333	.7155	.7972	.8784	.9590
5	1.5520	1.6347	1.7169	1.7986	1.8797	1.9603
6	.5534	.6361	.7183	.8000	.8811	.9617
7	.5547	.6374	.7196	.8013	.8824	.9630
8	.5561	.6388	.7210	.8027	.8838	.9643
9	.5575	.6402	.7224	.8040	.8851	.9657
10	1.5589	1.6416	1.7237	1.8054	1.8865	1.9670
11	.5603	.6429	.7251	.8067	.8878	.9683
12	.5616	.6443	.7265	.8081	.8892	.9697
13	.5630	.6457	.7292	.8095	.8906	.9711
14	.5644	.6471	.7242	.8108	.8919	.9724
15	1.5658	1.6484	1.7306	1.8122	1.8932	1.9737
16	.5672	.6498	.7319	.8135	.8946	.9750
17	.5686	.6512	.7333	.8149	.8959	.9764
18	.5699	.6525	.7347	.8162	.8973	.9777
19	.5713	.6539	.7360	.8176	.8986	.9790
20	1.5727	1.6553	1.7374	1.8189	1.8999	1.9804
21	.5741	.6567	.7387	.8203	.9013	.9817
22	.5755	.6580	.7401	.8217	.9026	.9830
23	.5768	.6594	.7415	.8230	.9040	.9844
24	.5782	.6608	.7428	.8244	.9053	.9857
25	1.5796	1.6622	1.7442	1.8257	1.9067	1.9870
26	.5810	.6635	.7456	.8271	.9080	.9884
27	.5824	.6649	.7469	.8284	.9094	.9897
28	.5837	.6663	.7483	.8298	.9107	.9911
29	.5851	.6676	.7496	.8311	.9120	.9924
30	1.5865	1.6690	1.7510	1.8325	1.9134	1.9937
31	.5879	.6704	.7524	.8338	.9147	.9951
32	.5893	.6718	.7537	.8352	.9161	.9964
33	.5906	.6731	.7551	.8365	.9174	.9977
34	.5920	.6745	.7565	.8379	.9188	.9991
35	1.5934	1.6759	1.7578	1.8392	1.9201	2.0004
36	.5948	.6772	.7592	.8406	.9215	.0017
37	.5961	.6786	.7605	.8419	.9228	.0031
38	.5975	.6800	.7619	.8433	.9241	.0044
39	.5989	.6813	.7633	.8447	.9255	.0057
40	1.6003	1.6827	1.7646	1.8460	1.9268	2.0070
41	.6017	.6841	.7660	.8474	.9282	.0084
42	.6030	.6855	.7673	.8487	.9295	.0097
43	.6044	.6868	.7687	.8501	.9308	.0110
44	.6058	.6882	.7701	.8514	.9322	.0124
45	1.6072	1.6896	1.7714	1.8528	1.9335	2.0137
46	.6085	.6909	.7728	.8541	.9349	.0150
47	.6099	.6923	.7742	.8555	.9362	.0164
48	.6113	.6937	.7755	.8568	.9376	.0177
49	.6127	.6950	.7769	.8582	.9389	.0190
50	1.6141	1.6964	1.7782	1.8595	1.9402	2.0204
51	.6154	.6978	.7796	.8609	.9416	.0217
52	.6168	.6991	.7809	.8622	.9429	.0230
53	.6182	.7005	.7823	.8636	.9443	.0244
54	.6196	.7019	.7837	.8649	.9456	.0257
55	1.6209	1.7032	1.7850	1.8663	1.9469	2.0270
56	.6223	.7046	.7864	.8676	.9483	.0283
57	.6237	.7060	.7877	.8690	.9496	.0297
58	.6251	.7073	.7891	.8703	.9510	.0310
59	.6264	.7087	.7905	.8717	.9523	.0323
60	1.6278	1.7101	1.7918	1.8730	1.9536	2.0337

TABLES OF CONSTANTS FOR 5-INCH SINE BAR

M	24°	25°	26°	27°	28°	29°
0	2.0337	2.1131	2.1918	2.2699	2.3473	2.4240
1	.0350	.1144	.1931	.2712	.3486	.4253
2	.0363	.1157	.1944	.2725	.3499	.4266
3	.0376	.1170	.1958	.2738	.3512	.4278
4	.0390	.1183	.1971	.2751	.3525	.4291
5	2.0403	2.1197	2.1984	2.2764	2.3538	2.4304
6	.0416	.1210	.1997	.2777	.3550	.4317
7	.0430	.1223	.2010	.2790	.3563	.4329
8	.0443	.1236	.2023	.2803	.3576	.4342
9	.0456	.1249	.2036	.2816	.3589	.4355
10	2.0469	2.1262	2.2049	2.2829	2.3602	2.4367
11	.0483	.1276	.2062	.2842	.3614	.4380
12	.0496	.1289	.2075	.2855	.3627	.4393
13	.0509	.1302	.2088	.2868	.3640	.4405
14	.0522	.1315	.2101	.2881	.3653	.4418
15	2.0536	2.1328	2.2114	2.2893	2.3666	2.4431
16	.0549	.1341	.2127	.2906	.3679	.4444
17	.0562	.1354	.2140	.2919	.3691	.4456
18	.0575	.1368	.2153	.2932	.3704	.4469
19	.0589	.1381	.2166	.2945	.3717	.4482
20	2.0602	2.1394	2.2179	2.2958	2.3730	2.4494
21	.0615	.1407	.2192	.2971	.3743	.4507
22	.0628	.1420	.2205	.2984	.3755	.4520
23	.0642	.1433	.2218	.2997	.3768	.4532
24	.0655	.1447	.2232	.3010	.3781	.4545
25	2.0668	2.1460	2.2245	2.3023	2.3794	2.4558
26	.0681	.1473	.2258	.3036	.3807	.4570
27	.0695	.1486	.2271	.3048	.3819	.4583
28	.0708	.1499	.2284	.3061	.3832	.4596
29	.0721	.1512	.2297	.3074	.3845	.4608
30	2.0734	2.1525	2.2310	2.3087	2.3858	2.4621
31	.0748	.1538	.2323	.3100	.3870	.4634
32	.0761	.1552	.2336	.3113	.3883	.4646
33	.0774	.1565	.2349	.3126	.3896	.4659
34	.0787	.1578	.2362	.3139	.3909	.4672
35	2.0801	2.1591	2.2375	2.3152	2.3922	2.4684
36	.0814	.1604	.2388	.3165	.3934	.4697
37	.0827	.1617	.2401	.3177	.3947	.4709
38	.0840	.1630	.2414	.3190	.3960	.4722
39	.0853	.1643	.2427	.3203	.3973	.4735
40	2.0867	2.1656	2.2440	2.3216	2.3985	2.4747
41	.0880	.1670	.2453	.3229	.3998	.4760
42	.0893	.1683	.2466	.3242	.4011	.4773
43	.0906	.1696	.2479	.3255	.4024	.4785
44	.0920	.1709	.2492	.3268	.4036	.4798
45	2.0933	2.1722	2.2505	2.3280	2.4049	2.4811
46	.0946	.1735	.2518	.3293	.4062	.4823
47	.0959	.1748	.2531	.3306	.4075	.4836
48	.0972	.1761	.2544	.3319	.4087	.4848
49	.0986	.1774	.2557	.3332	.4100	.4861
50	2.0999	2.1787	2.2570	2.3345	2.4113	2.4874
51	.1012	.1801	.2583	.3358	.4126	.4886
52	.1025	.1814	.2596	.3371	.4138	.4899
53	.1038	.1827	.2609	.3383	.4151	.4912
54	.1052	.1840	.2621	.3396	.4164	.4924
55	2.1065	2.1853	2.2634	2.3409	2.4177	2.4937
56	.1078	.1866	.2647	.3422	.4189	.4949
57	.1091	.1879	.2660	.3435	.4202	.4962
58	.1104	.1892	.2673	.3448	.4215	.4975
59	.1117	.1905	.2686	.3460	.4228	.4987
60	2.1131	2.1918	2.2699	2.3473	2.4240	2.5000

TABLES OF CONSTANTS FOR 5-INCH SINE BAR

M	30°	31°	32°	33°	34°	35°
0	2.5000	2.5752	2.6496	2.7232	2,7959	2.8679
1	.5012	.5764	.6508	.7244	.7971	.8690
2	.5025	.5777	.6520	.7256	.7984	.8702
3	.5038	.5789	.6533	.7268	.7996	.8714
4	.5050	.5802	.6545	.7280	.8008	.8726
5	2.5063	2.5814	2.6557	2.7293	2.8020	2.8738
6	.5075	.5826	.6570	.7305	.8032	.8750
7	.5088	.5839	.6582	.7317	.8044	.8762
8	.5100	.5851	.6594	.7329	.8056	.8774
9	.5113	.5864	.6607	.7341	.8068	.8786
10	2.5126	2.5876	2.6619	2.7354	2.8080	2.8798
11	.5138	.5889	.6631	.7366	.8092	.8809
12	.5151	.5901	.6644	.7378	.8104	.8821
13	.5163	.5914	.6656	.7390	.8116	.8833
14	.5176	.5926	.6668	.7402	.8128	.8845
15	2.5188	2.5938	2.6680	2.7414	2.8140	2.8857
16	.5201	.5951	.6693	.7427	.8152	.8869
17	.5214	.5963	.6705	.7439	.8164	.8881
18	.5226	.5976	.6717	.7451	.8176	.8893
19	.5239	.5988	.6730	.7463	.8188	.8905
20	2.5251	2.6001	2.6742	2.7475	2.8200	2.8916
21	.5264	.6013	.6754	.7487	.8212	.8928
22	.5276	.6025	.6767	.7499	.8224	.8940
23	.5289	.6038	.6779	.7512	.8236	.8952
24	.5301	.6050	.6791	.7524	.8248	.8964
25	2.5314	2.6063	2.6803	2.7536	2.8260	2.8976
26	.5327	.6075	.6816	.7548	.8272	.8988
27	.5339	.6087	.6828	.7560	.8284	.8999
28	.5352	.6100	.6840	.7572	.8296	.9011
29	.5364	.6112	.6852	.7584	.8308	.9023
30	2.5377	2.6125	2.6865	2.7597	2.8320	2.9035
31	.5389	.6137	.6877	.7609	.8332	.9047
32	.5402	.6149	.6889	.7621	.8344	.9059
33	.5414	.6162	.6902	.7633	.8356	.9070
34	.5427	.6174	.6914	.7645	.8368	.9082
35	2.5439	2.6187	2.6926	2.7657	2.8380	2.9094
36	.5452	.6199	.6938	.7669	.8392	.9106
37	.5464	.6211	.6951	.7681	.8404	.9118
38	.5477	.6224	.6963	.7694	.8416	.9130
39	.5489	.6236	.6975	.7706	.8428	.9141
40	2.5502	2.6249	2.6987	2.7718	2.8440	2.9153
41	.5514	.6261	.7000	.7730	.8452	.9165
42	.5527	.6273	.7012	.7742	.8464	.9177
43	.5539	.6286	.7024	.7754	.8476	.9189
44	.5552	.6298	.7036	.7766	.8488	.9200
45	2.5564	2.6310	2.7048	2.7778	2.8500	2.9212
46	.5577	.6323	.7061	.7790	.8512	.9224
47	.5589	.6335	.7073	.7802	.8523	.9236
48	.5602	.6348	.7085	.7815	.8535	.9248
49	.5614	.6360	.7097	.7827	.8547	.9259
50	2.5627	2.6372	2.7110	2.7839	2.8559	2.9271
51	.5639	.6385	.7122	.7851	.8571	.9283
52	.5652	.6397	.7134	.7863	.8583	.9295
53	.5664	.6409	.7146	.7875	.8595	.9307
54	.5677	.6422	.7158	.7887	.8607	.9318
55	2.5689	2.6434	2.7171	2.7899	2.8619	2.9330
56	.5702	.6446	.7183	.7911	.8631	.9342
57	.5714	.6459	.7195	.7923	.8643	.9354
58	.5727	.6471	.7207	.7935	.8655	.9365
59	.5739	.6483	.7220	.7947	.8667	.9377
60	2.5752	2.6496	2.7232	2.7959	2.8679	2.9389

TABLES OF CONSTANTS FOR 5-INCH SINE BAR

M	42°	43°	44°	45°	46°	47°
0	3.3456	3.4100	3.4733	3.5355	3.5967	3.6567
1	.3467	.4110	.4743	.5365	.5977	.6577
2	.3478	.4121	.4754	.5376	.5987	.6587
3	.3489	.4132	.4764	.5386	.5997	.6597
4	.3499	.4142	.4774	.5396	.6007	.6607
5	3.3510	3.4153	3.4785	3.5406	3.6017	3.6617
6	.3521	.4163	.4795	.5417	.6027	.6627
7	.3532	.4174	.4806	.5427	.6037	.6637
8	.3543	.4185	.4816	.5437	.6047	.6647
9	.3553	.4195	.4827	.5448	.6058	.6657
10	3.3564	3.4206	3.4837	3.5458	3.6068	3.6666
11	.3575	.4217	.4848	.5468	.6078	.6676
12	.3586	.4227	.4858	.5478	.6088	.6686
13	.3597	.4238	.4868	.5489	.6098	.6696
14	.3607	.4248	.4879	.5499	.6108	.6706
15	3.3618	3.4259	3.4889	3.5509	3.6118	3.6716
16	.3629	.4269	.4900	.5519	.6128	.6726
17	.3640	.4280	.4910	.5529	.6138	.6736
18	.3651	.4291	.4921	.5540	.6148	.6745
19	.3661	.4301	.4931	.5550	.6158	.6755
20	3.3672	3.4312	3.4941	3.5560	3.6168	3.6765
21	.3683	.4322	.4952	.5570	.6178	.6775
22	.3693	.4333	.4962	.5581	.6188	.6785
23	.3704	.4344	.4973	.5591	.6198	.6795
24	.3715	.4354	.4983	.5601	.6208	.6805
25	3.3726	3.4365	3.4993	3.5611	3.6218	3.6814
26	.3736	.4375	.5004	.5621	.6228	.6824
27	.3747	.4386	.5014	.5632	.6238	.6834
28	.3758	.4396	.5024	.5642	.6248	.6844
29	.3769	.4407	.5035	.5652	.6258	.6854
30	3.3779	3.4417	3.5045	3.5662	3.6268	3.6864
31	.3790	.4428	.5056	.5672	.6278	.6873
32	.3801	.4439	.5066	.5683	.6288	.6883
33	.3811	.4449	.5076	.5693	.6298	.6893
34	.3822	.4460	.5087	.5703	.6308	.6903
35	3.3833	3.4470	3.5097	3.5713	3.6318	3.6913
36	.3844	.4481	.5107	.5723	.6328	.6923
37	.3854	.4491	.5118	.5734	.6338	.6932
38	.3865	.4502	.5128	.5744	.6348	.6942
39	.3876	.4512	.5138	.5754	.6358	.6952
40	3.3886	3.4523	3.5149	3.5764	3.6368	3.6962
41	.3897	.4533	.5159	.5774	.6378	.6972
42	.3908	.4544	.5169	.5784	.6388	.6981
43	.3918	.4554	.5180	.5795	.6398	.6991
44	.3929	.4565	.5190	.5805	.6408	.7001
45	3.3940	3.4575	3.5200	3.5815	3.6418	3.7011
46	.3950	.4586	.5211	.5825	.6428	.7020
47	.3961	.4596	.5221	.5835	.6438	.7030
48	.3972	.4607	.5231	.5845	.6448	.7040
49	.3982	.4617	.5242	.5855	.6458	.7050
50	3.3993	3.4628	3.5252	3.5866	3.6468	3.7060
51	.4004	.4638	.5262	.5876	.6478	.7069
52	.4014	.4649	.5273	.5886	.6488	.7079
53	.4025	.4659	.5283	.5896	.6498	.7089
54	.4036	.4670	.5293	.5906	.6508	.7099
55	3.4046	3.4680	3.5304	3.5916	3.6518	3.7108
56	.4057	.4691	.5314	.5926	.6528	.7118
57	.4068	.4701	.5324	.5936	.6538	.7128
58	.4078	.4712	.5335	.5947	.6548	.7138
59	.4089	.4722	.5345	.5957	.6558	.7147
60	3.4100	3.4733	3.5355	3.5967	3.6567	3.7157

TABLES OF CONSTANTS FOR 5-INCH SINE BAR

M	36°	37°	38°	39°	40°	41°
0	2.9389	3.0091	3.0783	3.1466	3.2139	3.2803
1	.9401	.0102	.0794	.1477	.2150	.2814
2	.9413	.0114	.0806	.1488	.2161	.2825
3	.9424	.0125	.0817	.1500	.2173	.2836
4	.9436	.0137	.0829	.1511	.2184	.2847
5	2.9448	3.0149	3.0840	3.1522	3.2195	3.2858
6	.9460	.0160	.0852	.1534	.2206	.2869
7	.9471	.0172	.0863	.1545	.2217	.2879
8	.9483	.0183	.0874	.1556	.2228	.2890
9	.9495	.0195	.0886	.1567	.2239	.2901
10	2.9507	3.0207	3.0897	3.1579	3.2250	3.2912
11	.9518	.0218	.0909	.1590	.2262	.2923
12	.9530	.0230	.0920	.1601	.2273	.2934
13	.9542	.0241	.0932	.1612	.2284	.2945
14	.9554	.0253	.0943	.1624	.2295	.2956
15	2.9565	3.0264	3.0954	3.1635	3.2306	3.2967
16	.9577	.0276	.0966	.1646	.2317	.2978
17	.9589	.0288	.0977	.1658	.2328	.2989
18	.9600	.0299	.0989	.1669	.2339	.3000
19	.9612	.0311	.1000	.1680	.2350	.3011
20	2.9624	3.0322	3.1012	3.1691	3.2361	3.3022
21	.9636	.0334	.1023	.1703	.2373	.3033
22	.9647	.0345	.1034	.1714	.2384	.3044
23	.9659	.0357	.1046	.1725	.2395	.3054
24	.9671	.0369	.1057	.1736	.2406	.3065
25	2.9682	3.0380	3.1069	3.1748	3.2417	3.3076
26	.9694	.0392	.1080	.1759	.2428	.3087
27	.9706	.0403	.1091	.1770	.2439	.3098
28	.9718	.0415	.1103	.1781	.2450	.3109
29	.9729	.0426	.1114	.1792	.2461	.3120
30	2.9741	3.0438	3.1125	3.1804	3.2472	3.3131
31	.9753	0449	.1137	.1815	.2483	.3142
32	.9764	0461	.1148	.1826	.2494	.3153
33	.9776	.0472	.1160	.1837	.2505	.3163
34	.9788	.0484	.1171	.1849	.2516	.3174
35	2.9799	3.0495	3.1182	3.1860	3.2527	3.3185
36	.9811	.0507	.1194	.1871	.2538	.3196
37	.9823	.0519	.1205	.1882	.2550	.3207
38	.9834	.0530	.1216	.1893	.2561	.3218
39	.9846	.0542	.1228	.1905	.2572	.3229
40	2.9858	3.0553	3.1239	3.1916	3.2583	3.3240
41	.9869	.0565	.1251	.1927	.2594	.3250
42	.9881	.0576	.1262	.1938	.2605	.3261
43	.9893	.0588	.1273	.1949	.2616	.3272
44	.9904	.0599	.1285	.1961	.2627	.3283
45	2.9916	3.0611	3.1296	3.1972	3.2638	3.3294
46	.9928	.0622	.1307	.1983	.2649	.3305
47	.9939	.0634	.1319	.1994	.2660	.3316
48	.9951	.0645	.1330	.2005	.2671	.3326
49	.9963	.0657	.1341	.2016	.2682	.3337
50	2.9974	3.0668	3.1353	3.2028	3.2693	3.3348
51	.9986	.0680	.1364	.2039	.2704	.3359
52	.9997	.0691	.1375	.2050	.2715	.3370
53	3.0009	.0703	.1387	.2061	.2726	.3381
54	.0021	.0714	.1398	.2072	.2737	.3391
55	3.0032	3.0725	3.1409	3.2083	3.2748	3.3402
56	.0044	.0737	.1421	.2095	.2759	.3413
57	.0056	.0748	.1432	.2106	.2770	.3424
58	.0067	.0760	.1443	.2117	.2781	.3435
59	.0079	.0771	.1454	.2128	.2792	.3445
60	3.0091	3.0783	3.1466	3.2139	3.2803	3.3456

TABLES OF CONSTANTS FOR 5-INCH SINE BAR

M	48°	49°	50°	51°	52°	53°
0	3.7157	3.7735	3.8302	3.8857	3.9400	3.9932
1	.7167	.7745	.8311	.8866	.9409	.9940
2	.7176	.7754	.8321	.8875	.9418	.9949
3	.7186	.7764	.8330	.8884	.9427	.9958
4	.7196	.7773	.8339	.8894	.9436	.9967
5	3.7206	3.7783	3.8349	3.8903	3.9445	3.9975
6	.7215	.7792	.8358	.8912	.9454	.9984
7	.7225	.7802	.8367	.8921	.9463	.9993
8	.7235	.7811	.8377	.8930	.9472	4.0001
9	.7244	.7821	.8386	.8939	.9481	.0010
10	3.7254	3.7830	3.8395	3.8948	3.9490	4.0019
11	.7264	.7840	.8405	.8958	.9499	.0028
12	.7274	.7850	.8414	.8967	.9508	.0036
13	.7283	.7859	.8423	.8976	.9516	.0045
14	.7293	.7869	.8433	.8985	.9525	.0054
15	3.7303	3.7878	3.8442	3.8994	3.9534	4.0062
16	.7312	.7887	.8451	.9003	.9543	.0071
17	.7322	.7897	.8460	.9012	.9552	.0080
18	.7332	.7906	.8470	.9021	.9561	.0089
19	.7341	.7916	.8479	.9030	.9570	.0097
20	3.7351	3.7925	3.8488	3.9039	3.9579	4.0106
21	.7361	.7935	.8498	.9049	.9588	.0115
22	.7370	.7944	.8507	.9058	.9596	.0123
23	.7380	.7954	.8516	.9067	.9605	.0132
24	.7390	.7963	.8525	.9076	.9614	.0141
25	3.7399	3.7973	3.8535	3.9085	3.9623	4.0149
26	.7409	.7982	.8544	.9094	.9632	.0158
27	.7419	.7992	.8553	.9103	.9641	.0167
28	.7428	.8001	.8562	.9112	.9650	.0175
29	.7438	.8011	.8572	.9121	.9659	.0184
30	3.7448	3.8020	3.8581	3.9130	3.9667	4.0193
31	.7457	.8029	.8590	.9139	.9676	.0201
32	.7467	.8039	.8599	.9148	.9685	.0210
33	.7476	.8048	.8609	.9157	.9694	.0219
34	.7486	.8058	.8618	.9166	.9703	.0227
35	3.7496	3.8067	3.8627	3.9175	3.9712	4.0236
36	.7505	.8077	.8636	.9184	.9720	.0244
37	.7515	.8086	.8646	.9193	.9729	.0253
38	.7525	.8096	.8655	.9202	.9738	.0262
39	.7534	.8105	.8664	.9212	.9747	.0270
40	3.7544	3.8114	3.8673	3.9221	3.9756	4.0279
41	.7553	.8124	.8683	.9230	.9765	.0288
42	.7563	.8133	.8692	.9239	.9773	.0296
43	.7573	.8143	.8701	.9248	.9782	.0305
44	.7582	.8152	.8710	.9257	.9791	.0313
45	3.7592	3.8161	3.8719	3.9266	3.9800	4.0322
46	.7601	.8171	.8729	.9275	.9809	.0331
47	.7611	.8180	.8738	.9284	.9817	.0339
48	.7620	.8190	.8747	.9293	.9826	.0348
49	.7630	.8199	.8756	.9302	.9835	.0356
50	3.7640	3.8208	3.8765	3.9311	3.9844	4.0365
51	.7649	.8218	.8775	.9320	.9853	.0374
52	.7659	.8227	.8784	.9329	.9861	.0382
53	.7668	.8236	.8793	.9338	.9870	.0391
54	.7678	.8246	.8802	.9347	.9879	.0399
55	3.7687	3.8255	3.8811	3.9355	3.9886	4.0408
56	.7697	.8265	.8820	.9364	.9896	.0416
57	.7707	.8274	.8830	.9373	.9905	.0425
58	.7716	.8283	.8839	.9382	.9914	.0433
59	.7726	.8293	.8848	.9391	.9923	.0442
60	3.7735	3.8302	3.8857	3.9400	3.9932	4.0451

TABLES OF CONSTANTS FOR 5-INCH SINE BAR

M	54°	55°	56°	57°	58°	59°
0	4.0451	4.0957	4.1452	4.1933	4.2402	4.2858
1	.0459	.0966	.1460	.1941	.2410	.2866
2	.0468	.0974	.1468	.1949	.2418	.2873
3	.0476	.0982	.1476	.1957	.2425	.2881
4	.0485	.0991	.1484	.1965	.2433	.2888
5	4.0493	4.0999	4.1492	4.1973	4.2441	4.2896
6	.0502	.1007	.1500	.1981	.2448	.2903
7	.0510	.1016	.1508	.1989	.2456	.2910
8	.0519	.1024	.1517	.1997	.2464	.2913
9	.0527	.1032	.1525	.2004	.2471	.2925
10	4.0536	4.1041	4.1533	4.2012	4.2479	4.2933
11	.0544	.1049	.1541	.2020	.2487	.2940
12	.0553	.1057	.1549	.2028	.2494	.2948
13	.0561	.1066	.1557	.2036	.2502	.2955
14	.0570	.1074	.1565	.2044	.2510	.2963
15	4.0578	4.1082	4.1573	4.2052	4.2517	4.2970
16	.0587	.1090	.1581	.2060	.2525	.2978
17	.0595	.1099	.1589	.2067	.2533	.2985
18	.0604	.1107	.1597	.2075	.2540	.2992
19	.0612	.1115	.1606	.2083	.2548	.3000
20	4.0621	4.1124	4.1614	4.2091	4.2556	4.3007
21	.0629	.1132	.1622	.2099	.2563	.3015
22	.0638	.1140	.1630	.2107	.2571	.3022
23	.0646	.1148	.1638	.2115	.2578	.3029
24	.0655	.1157	.1646	.2122	.2586	.3037
25	4.0663	4.1165	4.1654	4.2130	4.2594	4.3044
26	.0672	.1173	.1662	.2138	.2601	.3052
27	.0680	.1181	.1670	.2146	.2609	.3059
28	.0689	.1190	.1678	.2154	.2617	.3066
29	.0697	.1198	.1686	.2162	.2624	.3074
30	4.0706	4.1206	4.1694	4.2169	4.2632	4.3081
31	.0714	.1214	.1702	.2177	.2639	.3089
32	.0722	.1223	.1710	.2185	.2647	.3096
33	.0731	.1231	.1718	.2193	.2655	.3103
34	.0739	.1239	.1726	.2201	.2662	.3111
35	4.0748	4.1247	4.1734	4.2208	4.2670	4.3118
36	.0756	.1255	.1742	.2216	.2677	.3125
37	.0765	.1264	.1750	.2224	.2685	.3133
38	.0773	.1272	.1758	.2232	.2692	.3140
39	.0781	.1280	.1766	.2240	.2700	.3147
40	4.0790	4.1288	4.1774	4.2247	4.2708	4.3155
41	.0798	.1296	.1782	.2255	.2715	.3162
42	.0807	.1305	.1790	.2263	.2723	.3170
43	.0815	.1313	.1798	.2271	.2730	.3177
44	.0823	.1321	.1806	.2278	.2738	.3184
45	4.0832	4.1329	4.1814	4.2286	4.2745	4.3192
46	.0840	.1337	.1822	.2294	.2753	.3199
47	.0849	.1346	.1830	.2302	.2760	.3206
48	.0857	.1354	.1838	.2309	.2768	.3213
49	.0865	.1362	.1846	.2317	.2775	.3221
50	4.0874	4.1370	4.1854	4.2325	4.2783	4.3228
51	.0882	.1378	.1862	.2333	.2791	.3235
52	.0891	.1386	.1870	.2340	.2798	.3243
53	.0899	.1395	.1878	.2348	.2806	.3250
54	.0907	.1403	.1886	.2356	.2813	.3257
55	4.0916	4.1411	4.1894	4.2364	4.2821	4.3265
56	.0924	.1419	.1902	.2371	.2828	.3272
57	.0932	.1427	.1909	.2379	.2836	.3279
58	.0941	.1435	.1917	.2387	.2843	.3286
59	.0949	.1443	.1925	.2394	.2851	.3294
60	4.0957	4.1452	4.1933	4.2402	4.2858	4.3301

WIRE SIZES FOR 60° THREADS

INCLUDING
UNIFIED AND AMERICAN SCREW THREADS
AMERICAN NATIONAL FORM THREADS
AMERICAN STANDARD PIPE THREADS

Pounds Pressure*	Threads Per Inch	Best Wire Size	P.D. = Measurement Over Wires	Pounds Pressure*	Threads Per Inch	Best Wire Size	P.D. = Measurement Over Wires
1	80	.00724	.01089	2½	16	.03608	.05411
1	72	.0082	.01203	2½	14	.04125	.06189
1	64	.00904	.01359	2½	13	.04441	.06664
1	56	.01031	.01547	2½	12	.04813	.07222
1	48	.01203	.01805	2½	11½	.05020	.07530
1	44	.01312	.01968	2½	11	.05251	.07880
1	40	.01444	.02167	2½	10	.05775	.08665
1	36	.01604	.02406	2½	9	.06415	.09623
1	32	.01804	.02706	2½	8	.07218	.10829
1	32	.01924	.02885	2½	7	.08248	.12372
1	28	.02062	.03093	2½	6	.09623	.14435
1	27	.02137	.03204	2½	5½	.10497	.15745
1	24	.02405	.03607	2½	5	.11548	.17323
2½	20	.02887	.04331	2½	4½	.12832	.19251
2½	18	.03207	.04810	2½	4	.14433	.21648

* Pounds Pressure Set on a Super Micrometer

COMMERCIAL TOLERANCES ON COILED SPRINGS*

Unless otherwise agreed, wire coil springs are furnished in accordance with the following variations:

Outer Diameter	+ or —
Less than 1/8 inch	0.003 inch
1/8 to 1/4 inch	0.005 inch
Over 1/4 to 1/2 inch	0.008 inch
Over 1/2 to 1 inch	1/64 inch
Over 1 to 2 inches	1/32 inch
Over 2 to 3 inches	1/16 inch
Over 3 to 4 inches	3/32 inch
Over 4 to 5 inches	7/64 inch
Over 5 to 6 inches	1/8 inch
Over 6 inches and up	1/8 inch

Length	+ or —
Less than 1 inch	1/32 inch
1 to 2 inches	1/16 inch
Over 2 to 3 inches	3/32 inch
Over 3 to 5 inches	1/8 inch
Over 5 to 8 inches	5/32 inch
Over 8 to 12 inches	1/4 inch
Over 12 to 18 inches	3/8 inch
Over 18 to 24 inches	1/2 inch
Over 24 to 30 inches	3/4 inch
Over 30 inches	1 inch

Number of Coils	or —+
Less than 4	1/4 Coil
Over 4 to 8	1/2
Over 8 to 15	3/4
Over 15 to 25	1

Squareness of Grinding For Index (OD÷d)	Degrees
Up to 6	3°
6 to 10	4°
Over 10 to 20	5°
Over 20	6°

Load, Rate and Deflection Tests + or — 10%
Position of Hooks + or — 45°
* These tolerances are for average spring indexes up to 10 inches.

GLOSSARY

ACTIVE COILS are those coils which are free to deflect when a load is applied to the spring. In most extension springs, all the coils are active. There is a definite difference between active coils and total coils in compression springs.

ALLOWANCE is the difference in the dimensions of mating parts, for example, the difference between the maximum diameter of a shaft and the minimum diameter of the mating bore.

ANGLE OF THREAD is the angle included between the sides of the thread measured in an axial plane.

ANVIL is a nonadjustable block of a micrometer or other gage.

AXIS OF A SCREW is the longitudinal central line through the screw.

BASE OF THREAD is the bottom section of the thread; the greatest section between the two adjacent roots.

BODY LENGTH refers to the over-all length of the active coils in an extension spring. This does not apply to compression springs, but is always a part of extension spring specifications.

CENTER POINT is a gage block attachment that is used to hold a piece of work on the center line.

COMPARATOR is a mechanism for amplifying and measuring a small displacement and is used to compare the unknown dimension of a part with some standard, such as with Hoke blocks.

CREST is the top surface joining the two sides of a thread.

DEGREE is an angle which is $\frac{1}{360}$th part of a circle.

DEPTH OF ENGAGEMENT is the depth of contact of two mating parts.

DEPTH OF THREAD is the distance between the crest and the base of thread measured normal to the axis.

DIMENSION is the number of units of length in the width, height, or length of an object.

DRIFT HOLE is a slot through which a drift is inserted for ejecting a part. The drift hole is sometimes called a drift slot.

DRILL ROD is a piece of steel with an outside diameter of close tolerance.

EXTERNAL THREAD is a thread on the outside of a part, such as a screw, stud or bolt.

FINISH is the structure, such as smoothness of a surface.

FIT is the dimensional relationship between two or more mating parts.

FLANGE is a rim for attachment to another part.

FLATNESS means that all points on the surface come within the specified limits as measured from a perfect plane.

FREE LENGTH is the length of a spring unloaded. This length is the over-all length of a compression spring or the length measured inside the hooks at the ends of extension springs.

GAGE is an instrument for determining whether or not the dimensions of a part are within the specified limits.

GAGING is a method of determining if a dimension of a part conforms to the specified size.

GO AND NO-GO GAGE is so constructed that the *go* section permits the part that is being measured to pass over the gaging surface, but prevents the part from passing over the *no-go* sections. The *go* and *no-go* surfaces are set within the specified limits.

HALF ANGLE OF THREAD is the angle included between a flank of the thread and the normal (90°) to the axis, measured in an axial plane.

HELIX ANGLE is the angle made by the helix of the thread at the pitch diameter with a plane perpendicular to the axis.

HERMAPHRODITE CALIPERS are used to layout for locating centers, laying out distances from an edge, etc.

INDICATING GAGE shows the variations in the dimensions of the part that is being measured. It is either a pointer or a shadow which travels over the face of a scale.

INSPECTION GAGES are gages used by an inspector to determine whether or not the part being inspected comes within the specifications.

INTERNAL THREAD is a thread on the inside of a part, such as the thread in a nut.

LEAD of a thread is the distance through which the screw will advance parallelly to the axis during one complete revolution.

LENGTH OF ENGAGEMENT is the length of contact between two mating parts, measured axially.

LIMITS are the maximum and minimum dimensions specified for a part.

MASTER GAGES are gages that are used for checking other gages.

MAJOR DIAMETER is the largest diameter of the thread of the screw or nut.

MEAN DIMENSION is the exact dimension desired and as it cannot be readily obtained, variations from the mean are specified.

MICRO INCH is one millionth (0.000001) part of the United States Standard inch (abbreviation = mu. in. or μ in.).

MICRON (mu. or μ) is the one thousandth part of one millimeter; the one millionth part of one meter.

MINOR DIAMETER is the smallest diameter of the thread of the screw or nut.

MINUS (—) is the variation allowed below the mean dimension.

MINUTE (′) is the $\frac{1}{60}$th part of an angle of 1°.

MONOCHROMATIC LIGHT is a light which gives off rays of one color.

NEUTRAL ZONE is a positive allowance between two mating parts.

NOMINAL SIZE is the named size which has no specified limits of accuracy.

NUMBER OF THREADS is the number of threads in 1 inch of length.

ORIFICE is an opening through which fluid or liquid may be discharged.

PARALLELISM indicates that the distance between the gaging surfaces is the same at all points.

PERMANENT SET occurs when a material is deflected so far that its elastic properties have been exceeded. When this is the case, a spring does not return to its original size on release of the load.

PITCH is the distance from a point on a screw thread to a corresponding point on the next thread measured parallelly to the axis.

PITCH DIAMETER on a straight screw thread is the diameter of an imaginary cylinder, the surface of which would pass through the threads at such points as to make equal the width of the threads and the width of the spaces cut by the surface of the cylinder. On a taper screw thread, it is the diameter, at a given distance from a reference plane perpendicular to the axis of an imaginary cone, the surface of which would pass through the threads at such points as to make equal the width of the threads and the width of the spaces cut by the surface of the cone.

PLUS is the variation allowed over the mean dimension.

REFERENCE GAGES are used for testing or checking inspection gages.

RING GAGE is a round gage used to check external diameters.

ROOT is the bottom surface joining the sides of two adjacent threads.

ROUGHNESS indicates minute contiguous irregularities occurring on the nominal surface of a plane.

SCREW JACK is a jack in which a screw is used for lifting.

SECOND (″) is a $\frac{1}{3600}$th part of a degree ($\frac{1}{60}$th of a minute).

SINE If two radii of a circle are drawn with the included angle less than a right angle and a perpendicular is erected from a point on one of the radii to the outer end of the other radius, then the ratio of the length of this perpendicular to the radius is called the sine of the included angle.

SPHERICAL DEPRESSION has the form of a sphere or one of its segments that is depressed; a hollow.

STANDARD is a physical representation of a dimension, size or form as established by law or by general usage and consent.

STANDARD GAGE is an exact value, or concept, that has been established to serve as a model or rule in the measurements of a quantity.

SUPER MICROMETER is a bench-type micrometer. The range of the micrometer head is 1 in. The measuring pressure control is in the tailstock and can be adjusted to give from 1 lb. to 2½ lb.

SURFACE FINISH is the smoothness of a surface as measured by surface measuring instruments, such as the profilometer.

THICKNESS OF THREAD is the distance between the adjacent sides of the thread measured along or parallel to the pitch line.

THREAD PLUG GAGE is a threaded cylindrical gage used for gaging internal threads.

THREAD RING GAGE is a ring, threaded internally, used for gaging the threads of an externally threaded part.

TOLERANCE is the variation above or below the mean dimension that is specified for the part being machined.

TOTAL COILS is the total number of coils in the spring, as the name implies. The term is used for compression springs only and need not be used if the active coils are specified.

TRAMMEL POINT is a narrow pointed block used to check the distance between lines or a center and lines.

WAVINESS is a deviation from the nominal surface evidenced by recurrent irregularities having the form of waves.

WIRE GAGE is a specification for the size of the wire to be used. Decimal specifications are desired since there are several gage tables causing some confusion as to the size desired.

WORKING GAGE is the gage that is used in production by the men producing the part.

WRING is the adherence of two gages having accurately flat surfaces when they are brought into contact.

STANDARD TABLES OF LIMITS

All holes are made to A.S.A. class IV limits.

The different classes of fits are obtained by varying the shaft size to get the required A.S.A. allowances.

CLASS I FIT (loose fit, large allowance, interchangeable)
This fit provides for considerable freedom and embraces certain fits where accuracy is not essential.

CLASS II FIT (free fit, liberal allowance, interchangeable)
For running speeds of 600 revolutions per minute or over and journal pressures of 600 pounds per square inch or over.

CLASS III FIT (medium fit, medium allowance, interchangeable)
For running fits under 600 revolutions per minute and with journal pressures less than 600 pounds per square inch also for sliding fits and more accurate machine tool parts.

CLASS IV FIT (snug fit, zero allowance, interchangeable)
This is the closest fit which can be assembled by hand and requires considerable precision. It should be used where no perceptible shake is permissible and where moving parts are not intended to move freely under load.

CLASS V FIT (wringing fit, zero to negative allowance, selective assembly)
This fit is practically metal to metal. Assembly is usually selective and not interchangeable.

CLASS VI FIT (tight fit, slight negative allowance, selective assembly)
Light pressure is required to assemble these fits and the parts are more or less permanently assembled, such as fixed ends of studs for gears, pulleys, rocker arms, etc.

CLASS VII FIT (medium force fit, negative allowance, selective assembly)
Considerable pressure is required to assemble these fits and the parts are considered permanently assembled. These fits are the tightest fits which are recommended for cast iron holes or external members as they stress cast iron to its elastic limit.

CLASS VIII FIT (heavy force and shrink fit, considerable negative allowance, selective assembly)
These fits are used for steel holes where metal can be stressed to its elastic limit. These fits cause excessive stress for cast iron holes.

INDEX

T

V

W